Novel Nano-Engineered Biomaterials for Bone Tissue Engineering

Novel Nano-Engineered Biomaterials for Bone Tissue Engineering

Editors

Sašo Ivanovski
Karan Gulati
Abdalla Abdal-hay

MDPI • Basel • Beijing • Wuhan • Barcelona • Belgrade • Manchester • Tokyo • Cluj • Tianjin

Editors
Sašo Ivanovski
The University of Queensland
Australia

Karan Gulati
The University of Queensland
Australia

Abdalla Abdal-hay
The University of Queensland
Australia

Editorial Office
MDPI
St. Alban-Anlage 66
4052 Basel, Switzerland

This is a reprint of articles from the Special Issue published online in the open access journal *Nanomaterials* (ISSN 2079-4991) (available at: https://www.mdpi.com/journal/nanomaterials/special_issues/nano_engineered_biomaterials).

For citation purposes, cite each article independently as indicated on the article page online and as indicated below:

LastName, A.A.; LastName, B.B.; LastName, C.C. Article Title. *Journal Name* **Year**, *Volume Number*, Page Range.

ISBN 978-3-0365-3349-0 (Hbk)
ISBN 978-3-0365-3350-6 (PDF)

Cover image courtesy of Karan Gulati

Contents

About the Editors

Sašo Ivanovski is the Dean of the School of Dentistry, Professor of Periodontology and director of the Centre for Orofacial Regeneration, Reconstruction and Rehabilitation (COR3) at the University of Queensland. He is a clinician-scientist and a board-certified specialist in periodontics, with a focus on surgical implant dentistry. Sašo is on the editorial board of multiple journals, including the Journal of Dental Research and Clinical Oral Implants Research. He is also currently president of the Australian and New Zealand Division of the International Association for Dental Research (IADR). Sašo is a previous federal president of the Australasian Osseointegration Society (AOS) and the Australian Society of Periodontology (ASP). He is a Fellow of the International College of Dentists (FICD), Academy of Dentistry International (FADI), Pierre Fauchard Academy (FPFA) and International Team of Implantology (ITI Senior Fellow). He has published over 200 papers in the peer-reviewed literature and leads an internationally recognized research group focusing on regenerative dentistry and tissue engineering, as well as clinical periodontics and implant dentistry.

Karan Gulati is a Research Group Leader and the Deputy Director of Research at the School of Dentistry, The University of Queensland. Dr Gulati is a pioneer in electrochemically nano-engineered implants with over 11 years of extensive research experience using nano-engineering towards various bioactive and therapeutic applications. He completed his PhD from the University of Adelaide in 2015 and was awarded the Dean's Commendation for Doctoral Thesis Excellence. His career has been supported by prestigious fellowships from NHMRC (National Health and Medical Research Council, Australia), JSPS (Japan Society for the Promotion of Science, Japan), Erasmus+ (Germany) and the University of Queensland.

Abdalla Abdal-hay is a biomaterials and tissue engineering expert researcher with a strong track record in the field. His research group (Ali3B) at the School of Dentistry, University of Queensland focuses on developing advanced biomaterials with multifunctional characteristics that meet the essential requirements for biomedical applications. Ali's research is focused on developing functional biomaterials, nanotechnology, additive manufacturing, and biodegradable metallic implants. In the last 10 years, Ali has published three book chapters and >70 publications in leading journals in dentistry and tissue engineering, as well as clinical periodontics and implant dentistry.

 *nanomaterials*

MDPI

Editorial

Novel Nano-Engineered Biomaterials for Bone Tissue Engineering

Karan Gulati [1,*], **Abdalla Abdal-hay** [1,2,*] **and Sašo Ivanovski** [1,*]

1   Centre for Orofacial Regeneration, Reconstruction and Rehabilitation (COR3), School of Dentistry,
    The University of Queensland, 288 Herston Road, Herston, QLD 4006, Australia
2   Department of Engineering Materials and Mechanical Design, Faculty of Engineering,
    South Valley University, Qena 83523, Egypt
*   Correspondence: k.gulati@uq.edu.au (K.G.); abdalla.ali@uq.edu.au (A.A.-h.); s.ivanovski@uq.edu.au (S.I.)

Citation: Gulati, K.; Abdal-hay, A.; Ivanovski, S. Novel Nano-Engineered Biomaterials for Bone Tissue Engineering. *Nanomaterials* **2022**, *12*, 333. https://doi.org/10.3390/nano12030333

Received: 18 January 2022
Accepted: 19 January 2022
Published: 21 January 2022

Publisher's Note: MDPI stays neutral with regard to jurisdictional claims in published maps and institutional affiliations.

This Special Issue of *Nanomaterials* explores the recent advances relating to nano-engineered strategies for biomaterials and implants in bone tissue engineering. Spanning across the orthopaedic and dental settings, nano-engineered biomaterials surpass the bone regeneration and integration performance of conventional macro- and micro-scaled biomaterials [1–3]. Resorbable biomaterials, including extracellular matrix (ECM)-mimicking nanofibrous scaffolds fabficated from natural or synthetic sources [4], and nanoscale modified non-resorbable metal-based implants [5], have shown promising outcomes, as demonstrated in various investigations [6]. The enabling effective local therapeutic action without any potential cytotoxicity, while maintaining clinical translatability has further advanced the fabrication of highly bioactive biomaterials, addressing challenges associated with conventional counterparts [7].

The aim of this Special Issue is to highlight key nano-engineering attempts that challenge current clinical standards and advance the domain of bone tissue engineering. This multidisciplinary Special Issue will inform the readers of the future prospects in this domain, bridging the gap between research and clinical translation. To this end, leading scientists across the globe have contributed a total of eight original research, communication-style research and review papers, presenting novel nano-engineering strategies that highlight advances in bone tissue engineering.

Hydroxyapatite (HA) is biocompatible and non-immunogenic, and has been widely applied as a scaffold or towards surface modification for biomaterials/implants [8]. In an extensive review, Fu et al. detailed inorganic nanomaterial-based therapy with a focus on Ca-P compounds (nano-hydroxyapatites), nano-silica and metallic nanomaterials (Ti, Mg, Zn, Au and alloys) that advance biomineralization and bone defect repairing [9]. Next, Dumitrescu et al. compared the fabrication, characterization and in vitro performance of nano-hydroxyapatite and xenografts [10]. In this study, powder synthesized from egg shells and treated with a microwave-assisted hydrothermal technique (HA1), alongside two commercial xenograft powders, Bio-Oss® and Gen-Os®, were characterized, and the results revealed that the surface of the HA1 nanoparticles and internal mesopores contributed to augmented biocompatibility and osteoconduction abilities. Further advancing nano-HA research, Cestari et al. reported on the extraction of HA nano-powders from cuttlefish bones, mussel shells, chicken eggshells and bioinspired amorphous calcium carbonate, which were consolidated into cylindrical pellets (via uniaxial pressing and sintering) [11]. Characterizations involving SEM, XRD, ICP/OES and in vitro cytotoxicity evaluations confirmed that phase composition depends on the Ca/P ratio and the HA source, and that cellular functions were influenced (with the best cell adhesion for 900 °C sintered egg shell-derived nano-HA). The abovementioned studies highlighted that biomimetic and biogenic HA-based nanomaterials hold great promise for bone regeneration applications.

While biomimetic HA may be effective at promoting osteogenesis, mechanical strength and tissue volume remain challenging. To address this, Choi et al. explored the fabrication of a composite hydrogel using biphasic calcium phosphate (BCP) and gelatin methacrylate (GelMA), and performed preosteoblast proliferation and differentiation assessment in vitro [12]. The findings showed that the composite maintained the volume/shape of the hydrogel and upregulated cell viability and bone differentiation ability.

It is known that the inability to achieve high purity and a lack of appropriate fabrication techniques may limit the application of synthetic HA in bone tissue engineering. Lim et al. reported on the synthesis of human teeth-derived bioceramics and tested its bone regeneration potential in mice calvarial defects in vivo [13]. The bioceramics showed no adverse effects (WST-1 assay) and favourable adhesion of human alveolar bone marrow stem cells. Further, when compared to controls, the bioceramic-treated defects in mice calvaria demonstrated augmented vascularization, demonstrating the potential of such biomaterials in bone regeneration.

It is well established that 3D fibrous scaffolds enable cellular interaction via nanoscale focal adhesion complexes [14]; however, how osteoblasts respond to such scaffolds and their downstream events remains unexplored. To this end, Han et al. assessed primary human osteoblast sensing and responses to polycaprolactone (PCL) fibrous 3D scaffolds (both aligned and random oriented, fabricated via melt electrowriting (MEW)) [15]. The authors reported that 3D scaffolds caused immature vinculin focal adhesion formation and significantly reduced the nuclear localization of the mechanosensor-yes-associated protein (YAP). Further, compared to random fibers, aligned fibers elongated the cell and nucleus shape and activated global DNA methylation.

With favourable biocompatibility, corrosion-resistance, biomechanics and ease of modification, titanium (Ti) is the most popular material choice for orthopaedic and dental implants [16,17]. However, bare Ti-based implants are bioinert, and in compromised patient conditions (poor bone quality/quantity), enhanced bioactivity performance is needed to orchestrate osteogenesis [18]. As a result, surface modification of Ti-based implants using various topographical, chemical, biological and therapeutic strategies in the macro-, micro- and nano-scales have been performed [19]. Among these, nano-engineered Ti implants have outperformed macro-and micro-scale implants by achieving appropriate bioactivity and therapeutic performance [20–23].

Agour et al. modified Ti implants by dip-coating layers of polyurethane (PU) with HA nanoparticles (NPs) and magnesium (Mg) particles onto alkali-treated Ti implants [24]. Surface characterization and cell response evaluation (MC3T3-E1 osteoblast-like cells) in vitro revealed that HA incorporation increased interfacial bonding (between the coating and Ti) and Mg/HA particles augmented cellular functions, including adhesion, proliferation and differentiation. The research confirms the influence of incorporating bioactive agents, such as HA and Mg particles, onto Ti implants for inducing bone formation.

In an interesting study, Otte et al. [25] aimed to address the challenges associated with Ti and its alloys as biomedical implants, including high cost, low hardness, poor wear properties and potential side effects associated with Ti ion leaching [26]. The authors developed a TiB nanowhisker-reinforced Ti composite with augmented mechanical characteristics, including hardness for appropriate biomedical performance. Briefly, a cost-effective TiB-reinforced alpha Ti matrix composite (TMC) was developed with the composite microstructure incorporating ultrahigh aspect TiB nanowhiskers. The TMC characterization revealed an increase of 304%, 170% and 180% in hardness, modulus and hardness to modulus ratio, respectively. This confirmed the excellent mechanical performance of TMC and its potential in biomedical implant applications.

To summarize, this Special Issue in *Nanomaterials*, entitled "Novel Nano-Engineered Biomaterials for Bone Tissue Engineering", showcases the latest nano-engineering research challenging the current standards in bone tissue engineering. Readers of this Special Issue will understand the current nanotechnology advances, clinical translation challenges and future prospects, encompassing both resorbable polymeric and non-resorbable metallic biomaterials, that enable controlled and tailored orchestration of osteogenesis at the biomaterial–bone interface. The editors thank all the contributing authors and reviewers who have contributed to the success of the Special Issue.

Funding: Karan Gulati is supported by the National Health and Medical Research Council (NHMRC) Early Career Fellowship (APP1140699).

Conflicts of Interest: The authors declare no conflict of interest.

References

1. Casanellas, I.; García-Lizarribar, A.; Lagunas, A.; Samitier, J. Producing 3D Biomimetic Nanomaterials for Musculoskeletal System Regeneration. *Front. Bioeng. Biotechnol.* **2018**, *6*, 128. [CrossRef]
2. Cross, L.M.; Thakur, A.; Jalili, N.A.; Detamore, M.; Gaharwar, A.K. Nanoengineered biomaterials for repair and regeneration of orthopedic tissue interfaces. *Acta Biomater.* **2016**, *42*, 2–17. [CrossRef]
3. Sofi, H.S.; Abdal-Hay, A.; Ivanovski, S.; Zhang, Y.S.; Sheikh, F.A. Electrospun nanofibers for the delivery of active drugs through nasal, oral and vaginal mucosa: Current status and future perspectives. *Mater. Sci. Eng. C* **2020**, *111*, 110756. [CrossRef]
4. Wade, R.J.; Burdick, J.A. Advances in nanofibrous scaffolds for biomedical applications: From electrospinning to self-assembly. *Nano Today* **2014**, *9*, 722–742. [CrossRef]
5. Martinez-Marquez, D.; Gulati, K.; Carty, C.P.; Stewart, R.A.; Ivanovski, S. Determining the relative importance of titania nanotubes characteristics on bone implant surface performance: A quality by design study with a fuzzy approach. *Mater. Sci. Eng. C* **2020**, *114*, 110995. [CrossRef] [PubMed]
6. Chopra, D.; Jayasree, A.; Guo, T.; Gulati, K.; Ivanovski, S. Advancing dental implants: Bioactive and therapeutic modifications of zirconia. *Bioact. Mater.* **2021**. [CrossRef]
7. Gulati, K.; Zhang, Y.; Di, P.; Liu, Y.; Ivanovski, S. Research to Clinics: Clinical Translation Considerations for Anodized Nano-Engineered Titanium Implants. *ACS Biomater. Sci. Eng.* **2021**. [CrossRef]
8. Zhang, Y.; Gulati, K.; Li, Z.; Di, P.; Liu, Y. Dental Implant Nano-Engineering: Advances, Limitations and Future Directions. *Nanomater.* **2021**, *11*, 2489. [CrossRef] [PubMed]
9. Fu, Y.; Cui, S.; Luo, D.; Liu, Y. Novel Inorganic Nanomaterial-Based Therapy for Bone Tissue Regeneration. *Nanomaterials* **2021**, *11*, 789. [CrossRef] [PubMed]
10. Dumitrescu, C.R.; Neacsu, I.A.; Surdu, V.A.; Nicoara, A.I.; Iordache, F.; Trusca, R.; Ciocan, L.T.; Ficai, A.; Andronescu, E. Nano-Hydroxyapatite vs. Xenografts: Synthesis, Characterization, and In Vitro Behavior. *Nanomaterials* **2021**, *11*, 2289. [CrossRef]
11. Cestari, F.; Agostinacchio, F.; Galotta, A.; Chemello, G.; Motta, A.; Sglavo, V. Nano-Hydroxyapatite Derived from Biogenic and Bioinspired Calcium Carbonates: Synthesis and In Vitro Bioactivity. *Nanomaterials* **2021**, *11*, 264. [CrossRef] [PubMed]
12. Choi, J.-B.; Kim, Y.-K.; Byeon, S.-M.; Park, J.-E.; Bae, T.-S.; Jang, Y.-S.; Lee, M.-H. Fabrication and Characterization of Biodegradable Gelatin Methacrylate/Biphasic Calcium Phosphate Composite Hydrogel for Bone Tissue Engineering. *Nanomaterials* **2021**, *11*, 617. [CrossRef]
13. Lim, K.-T.; Patel, D.K.; Dutta, S.D.; Choung, H.-W.; Jin, H.; Bhattacharjee, A.; Chung, J.H. Human Teeth-Derived Bioceramics for Improved Bone Regeneration. *Nanomaterials* **2020**, *10*, 2396. [CrossRef]
14. Abbasi, N.; Ivanovski, S.; Gulati, K.; Love, R.M.; Hamlet, S. Role of offset and gradient architectures of 3-D melt electrowritten scaffold on differentiation and mineralization of osteoblasts. *Biomater. Res.* **2020**, *24*, 1–16. [CrossRef] [PubMed]
15. Han, P.; Vaquette, C.; Abdal-Hay, A.; Ivanovski, S. The Mechanosensing and Global DNA Methylation of Human Osteoblasts on MEW Fibers. *Nanomaterials* **2021**, *11*, 2943. [CrossRef]
16. Chopra, D.; Gulati, K.; Ivanovski, S. Understanding and optimizing the antibacterial functions of anodized nano-engineered titanium implants. *Acta Biomater.* **2021**, *127*, 80–101. [CrossRef]
17. Guo, T.; Gulati, K.; Arora, H.; Han, P.; Fournier, B.; Ivanovski, S. Orchestrating soft tissue integration at the transmucosal region of titanium implants. *Acta Biomater.* **2021**, *124*, 33–49. [CrossRef] [PubMed]
18. Guo, T.; Gulati, K.; Arora, H.; Han, P.; Fournier, B.; Ivanovski, S. Race to invade: Understanding soft tissue integration at the transmucosal region of titanium dental implants. *Dent. Mater.* **2021**, *37*, 816–831. [CrossRef]
19. Gulati, K.; Kogawa, M.; Maher, S.; Atkins, G.; Findlay, D.; Losic, D. Titania Nanotubes for Local Drug Delivery from Implant Surfaces. In *Electrochemically Engineered Nanoporous Materials: Methods, Properties and Applications*; Losic, D., Santos, A., Eds.; Springer International Publishing: Cham, Switzerland, 2015; pp. 307–355. [CrossRef]
20. Gulati, K.; Moon, H.-J.; Kumar, P.S.; Han, P.; Ivanovski, S. Anodized anisotropic titanium surfaces for enhanced guidance of gingival fibroblasts. *Mater. Sci. Eng. C* **2020**, *112*, 110860. [CrossRef]
21. Rahman, S.; Gulati, K.; Kogawa, M.; Atkins, G.J.; Pivonka, P.; Findlay, D.M.; Losic, D. Drug diffusion, integration, and stability of nanoengineered drug-releasing implants in bone ex-vivo. *J. Biomed. Mater. Res. Part A* **2015**, *104*, 714–725. [CrossRef] [PubMed]
22. Abdal-Hay, A.; Gulati, K.; Fernandez-Medina, T.; Qian, M.; Ivanovski, S. In situ hydrothermal transformation of titanium surface into lithium-doped continuous nanowire network towards augmented bioactivity. *Appl. Surf. Sci.* **2020**, *505*, 144604. [CrossRef]
23. Chopra, D.; Gulati, K.; Ivanovski, S. Bed of nails: Bioinspired nano-texturing towards antibacterial and bioactivity functions. *Mater. Today Adv.* **2021**, *12*, 100176. [CrossRef]

24. Agour, M.; Abdal-Hay, A.; Hassan, M.; Bartnikowski, M.; Ivanovski, S. Alkali-Treated Titanium Coated with a Polyurethane, Magnesium and Hydroxyapatite Composite for Bone Tissue Engineering. *Nanomaterials* **2021**, *11*, 1129. [CrossRef] [PubMed]
25. Otte, J.A.; Zou, J.; Patel, R.; Lu, M.; Dargusch, M.S. TiB Nanowhisker Reinforced Titanium Matrix Composite with Improved Hardness for Biomedical Applications. *Nanomaterials* **2020**, *10*, 2480. [CrossRef] [PubMed]
26. Gulati, K.; Scimeca, J.-C.; Ivanovski, S.; Verron, E. Double-edged sword: Therapeutic efficacy versus toxicity evaluations of doped titanium implants. *Drug Discov. Today* **2021**, *26*, 2734–2742. [CrossRef] [PubMed]

Review

Novel Inorganic Nanomaterial-Based Therapy for Bone Tissue Regeneration

Yu Fu [1], Shengjie Cui [2], Dan Luo [3,*] and Yan Liu [2,*]

[1] Fourth Clinical Division, Peking University School and Hospital of Stomatology; National Engineering Laboratory for Digital and Material Technology of Stomatology, Beijing Key Laboratory of Digital Stomatology, Beijing 100081, China; fuyu19880715@126.com

[2] Laboratory of Biomimetic Nanomaterials, Department of Orthodontics, Peking University School and Hospital of Stomatology, National Engineering Laboratory for Digital and Material Technology of Stomatology; Beijing Key Laboratory of Digital Stomatology, Beijing 100081, China; cuishengjie@pku.edu.cn

[3] CAS Center for Excellence in Nanoscience, Beijing Key Laboratory of Micro-nano Energy and Sensor, Beijing Institute of Nanoenergy and Nanosystems, Chinese Academy of Sciences, Beijing 100083, China

* Correspondence: luodan@binn.cas.cn (D.L.); orthoyan@bjmu.edu.cn (Y.L.)

Abstract: Extensive bone defect repair remains a clinical challenge, since ideal implantable scaffolds require the integration of excellent biocompatibility, sufficient mechanical strength and high biological activity to support bone regeneration. The inorganic nanomaterial-based therapy is of great significance due to their excellent mechanical properties, adjustable biological interface and diversified functions. Calcium–phosphorus compounds, silica and metal-based materials are the most common categories of inorganic nanomaterials for bone defect repairing. Nano hydroxyapatites, similar to natural bone apatite minerals in terms of physiochemical and biological activities, are the most widely studied in the field of biomineralization. Nano silica could realize the bone-like hierarchical structure through biosilica mineralization process, and biomimetic silicifications could stimulate osteoblast activity for bone formation and also inhibit osteoclast differentiation. Novel metallic nanomaterials, including Ti, Mg, Zn and alloys, possess remarkable strength and stress absorption capacity, which could overcome the drawbacks of low mechanical properties of polymer-based materials and the brittleness of bioceramics. Moreover, the biodegradability, antibacterial activity and stem cell inducibility of metal nanomaterials can promote bone regeneration. In this review, the advantages of the novel inorganic nanomaterial-based therapy are summarized, laying the foundation for the development of novel bone regeneration strategies in future.

Keywords: inorganic nanomaterials; bone regeneration; nano hydroxyapatites; nano silica; metallic nanomaterials

Citation: Fu, Y.; Cui, S.; Luo, D.; Liu, Y. Novel Inorganic Nanomaterial-Based Therapy for Bone Tissue Regeneration. *Nanomaterials* **2021**, *11*, 789. https://doi.org/10.3390/nano11030789

Academic Editor: Maria J. Blanco Prieto

Received: 15 December 2020
Accepted: 16 March 2021
Published: 19 March 2021

1. Introduction

Bone consists of inorganic minerals and organic matrix, and its highly hierarchical structure ensures excellent mechanical properties to withstand stress [1]. Although bone can heal itself through dynamic remodeling when suffering minor damage, bone lesions of critical size that cannot be cured spontaneously are quite common in daily life [2]. Until now, autologous transplantation is still the gold standard for the treatment of large bone defects [3]. However, the high incidence of donor sites mobility and the limited volume of autologous bone grafts limit the large-scale promotion of bone replacement surgery [4]. The ideal bone graft substitutes should not only imitate the extracellular matrix (ECM) of natural bone to achieve excellent biocompatibility, more importantly, they must provide strong mechanical support for the defect tissue. Inorganic nanomaterials possess better mechanical strength than natural and synthetic polymer scaffolds and can maintain stability for several weeks in vivo to support the bone healing process in the early stage of regeneration, making them the most promising candidate for bone graft substitutes. Calcium–phosphorus

Nanomaterials **2021**, *11*, 789. https://doi.org/10.3390/nano11030789　　　　https://www.mdpi.com/journal/nanomaterials

compounds, silica and metal-based materials are the most common categories of inorganic nanomaterials for bone defect repairing. Among them, nano hydroxyapatites (nHAs) are widely studied due to their high similarity with natural bone apatite [5]. Nano silica has been proven to establish hierarchical structure and promote bone regeneration through a biosilicification process [6]. For metal materials, traditional bulk metal scaffolds have been used for permanent and temporary orthopedic applications [7]. Recently, biodegradable metallic nanomaterials are able to be manufactured through a 3D printing technology, namely additive manufacturing (AM), to produce personalized orthopedic implants [8]. According to computer-aided design data, the mechanical properties, pore size, porosity and surface characteristics of the implant can be perfectly controlled [9]. In this review, the advantages and applications of the above three kinds of inorganic nanomaterials in bone tissue engineering will be reviewed (Figure 1).

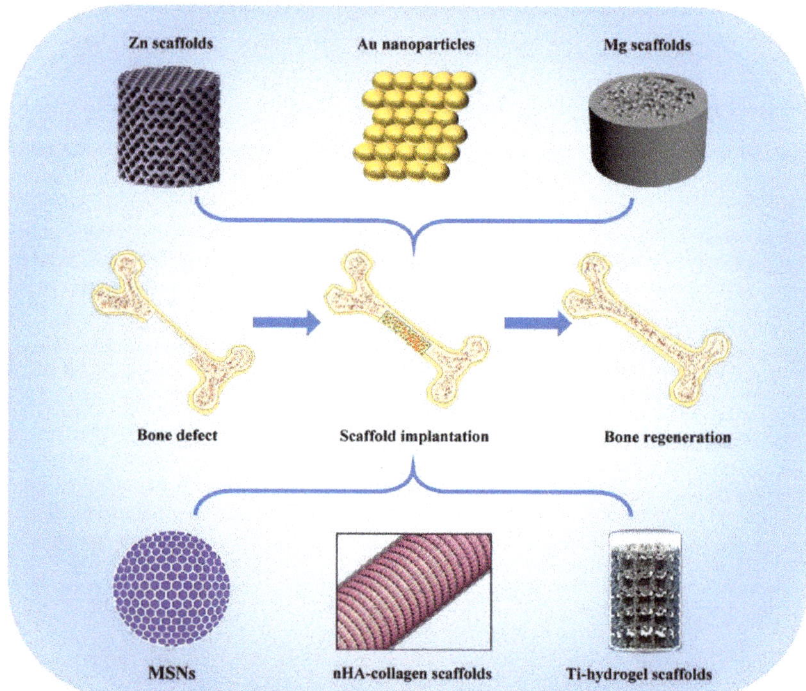

Figure 1. The main inorganic nanomaterials used for bone tissue regeneration [10–14].

2. Nano Hydroxyapatites

Natural mature bone is composed of minerals, type I collagen, water, small amounts of other collagen types, noncollagenous proteins and proteoglycans. The minerals mainly refer to thin plate-shaped carbonated calcium-deficient hydroxyapatites (50 nm × 25 nm in size, 1.5–4 nm thick), distributed along the collagen fibrils [1]. Hydroxyapatite (HA) has a Ca/P ratio of 1.67, higher than that of calcium-deficient hydroxyapatites, and is insoluble in vivo. HA is hard to degrade and has favorable mechanical properties and chemical binding ability, thus it is mostly used as a nanofiller in polymers to improve the mechanical strength or coating onto metallic implants to impart bioactivity [15].

2.1. nHA/Polymer Composites

Inspired by the composition and structure of bone, researchers try to mineralize organic matrix with calcium phosphates, especially nHAs, which are the most stable and similar member with natural mineral phase in vivo. It has been shown that mineralization with nHAs enhanced Young's modulus of type I collagen from 0.2–7.8 GPa to 11.1–16.3 GPa (Figure 2a) [16]. In addition to the improvement of mechanical properties, the mineralized fibrils showed rough surface topography and could promote the osteogenesis of mesenchymal stem cells (MSCs) in vitro and bone defect regeneration in vivo [17,18]. Even though chitosan had much higher compressive modulus (around 27 kPa) and compressive strength (around 36 kPa) compared to those of collagen (around 2 kPa and 3 kPa, respectively), nHAs were still able to significantly increase the mechanical properties of chitosan with compressive modulus of around 43 kPa and compressive strength of 40 kPa, respectively [19]. In another study, the Young's modulus of pure chitosan was reported to be approximately 3.0 GPa, while that of nHA/chitosan composite increased to over 3.5 GPa [20]. Similarly, the addition of nHAs could elevate the ultimate compressive strength of chitosan both in wet and dry conditions and the fabrication methods of nHA/chitosan composites further affected the mechanical strength [21]. Silk fibroin is also a natural polymer with compressive modulus of 0.43 MPa, and after being assembled with nHAs, its compressive modulus became 4 times higher (Figure 2b) [22]. Aside from the abovementioned natural polymers, the synthetic polymers also need nHAs in the fabrication process. It has been shown that HA nanorods can increase the elastic modulus and strength of polycaprolactone by approximately 50% and 26%, respectively [23]. However, there was an exception when nHAs were used to fabricate nHA/poly (ester urea) composite scaffold through 3D printing technology. The addition of nHAs had no significant influence on compressive modulus, which may result from the balance between the reinforcement effect and nonoptimized nHA/poly(ester urea) interaction [24]. Unlike the controversial effect on polymers' mechanical performance, nHAs can delay the degradation of poly(D, L-lactic acid)/poly(D, L-lactic- co-glycolic acid) blends [25]. Impressively, the effect of nHAs on the degradation rate of chitosan was even higher than that of nanobioglass, which may be attributed to the fine dispersion of nHAs in composites [26]. The addition of nHAs not only affects the mechanical performance and degradation rate of organic polymers, but also releases Ca^{2+} and PO_4^{3-} ions and regulates osteogenesis and bone regeneration [27]. For instance, the appearance of nHAs in poly(ester urea) scaffold significantly increased the alkaline phosphatase (ALP) activity, bone sialoprotein (BSP) and osteocalcin (OCN) expression and calcium deposition of MC3T3-E1 preosteoblast cells (Figure 2d) [24]. Gonzalez Ocampo et al. also proved nHAs promoted the cell spreading, ALP activity and matrix mineralization of human osteoblasts seeded on kappa-carrageenan (Figure 2c) [28]. Furthermore, nHA/polymer composites can promote bone regeneration in both calvarial defect (Figure 2e) [22] and long bone defect models [29]. Based on the inspiration from cortical bone and nacre, Feng et al. fabricated a "brick and mortar" multilayer nHA-based scaffold, which induced not only osteogenesis, but excellent angiogenesis as well (Figure 2f,g) [29]. The effect of nHAs coating on metal-based nanomaterials will be briefly discussed in Section 4.

2.2. 3D Printed nHA-Based Inorganic Nanomaterials

The latest emerging 3D printing technology makes nHAs and other inorganic nanomaterials easy to shape and may help to solve the poor biodegradability and excessive mechanical properties of biomaterial scaffolds consisting only or mainly of nHAs. Compared with the commercially available particle-type bone substitutes OSTEON 3 (Genoss®), the 3D printed customizable HA/tricalcium phosphate scaffolds promoted more new bone formation (Figure 3a) [30]. Even using the same raw materials calcium phosphate cement and polyvinyl butyral, different solvent of polyvinyl butyral, such as ethanol and tetrahydrofuran could lead to quite different geometry, microstructure, mechanical properties and osteoconductivity [31]. nHA/polyvinyl butyral composite scaffold fabricated in ethanol had 2-fold higher ultimate tensile strength and 3.4-fold higher ultimate compressive strength and promoted the osteogenesis of human primary osteoblasts, compared with that made in tetrahydrofuran. In addition to internal structure, surface topography and chemical characteristics also contribute to the bone reparation effect. Wei et al. reported that the hexagon-like column array topography of 3D printed HA scaffold promoted the osteogenic differentiation of human adipose-derived stem cells [32]. Furthermore, this research group found that the surface modification with strontium ion substitution can enhance the effect of HA porous scaffold on osteogenesis [33]. Similarly, the ECM derived from bone marrow MSCs (BMSCs) modified the surface chemistry of 3D HA scaffold and then improved the ALP activity, osteogenesis-related mRNA expression and calcium deposition of BMSCs, which finally promoted the bone reparation in rat skull defects (Figure 3b) [34]. As for the pore size, the HA-based 3D printed scaffolds with 1.4 mm and 1.2 mm pore sizes can promote bone regeneration at 4 weeks in rabbit calvarial defects while decreasing the mechanical strength, compared with those scaffolds with 0.8 mm and 1.0 mm pore sizes. However, the effect of pore size on bone regeneration was diminished as time went on to 8 weeks (Figure 3c) [35]. Future research is required to explore better methods to accurately determine the topography of nHA-based 3D printing scaffolds.

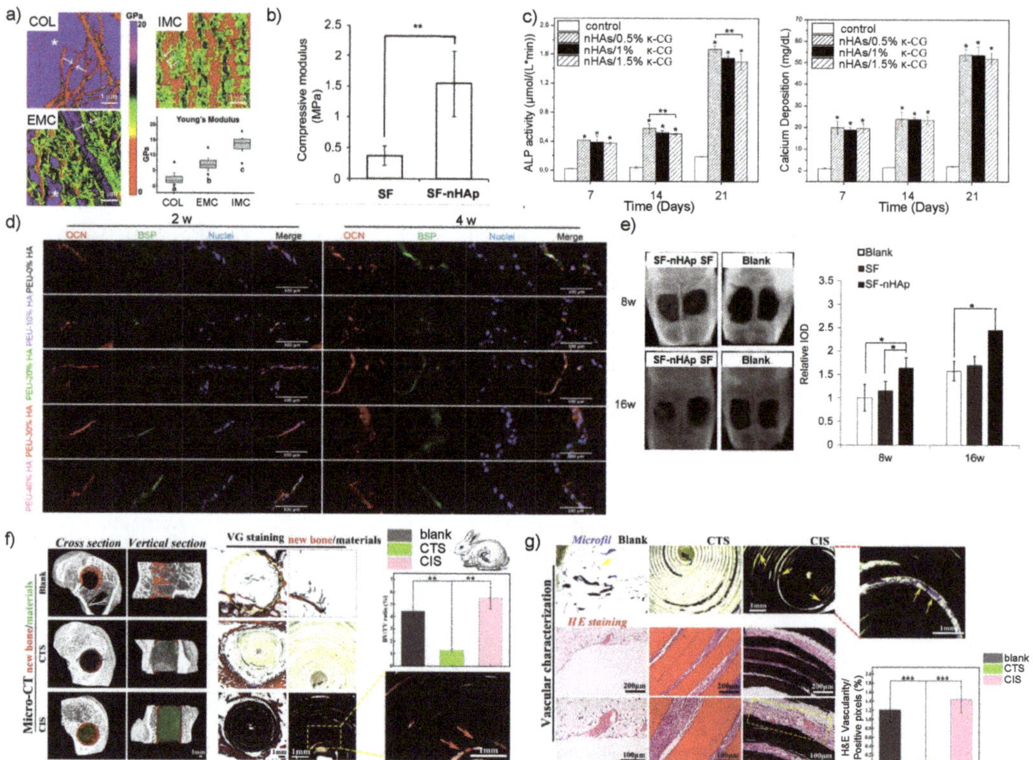

Figure 2. The mechanical and biological properties of nHA/polymer composites. (**a**) The highly ordered deposited nano hydroxyapatites (nHAs) provided intrafibrillarly mineralized collagen (IMC) with much higher Young's modulus than that of pure collagen (COL) and extrafibrillarly mineralized collagen (EMC) [16]. (**b**) The addition of nHAs increased the compressive modulus of silk fibroin significantly [22]. (**c**) nHA/kappa-carrageenan (κ-CG) enhanced the alkaline phosphatase (ALP) activity and calcium deposition of human osteoblasts, regardless of the concentration of κ-CG [28]. (**d**) With the increase of nHAs content, the MC3T3-E1 preosteoblast cells expressed more osteocalcin (OCN) and bone sialoprotein (BSP) [24]. (**e**) The nHA/silk fibroin composites promoted the bone regeneration in rat calvarial defects [22]. The co-inspired scaffold (CIS) which consisted of nHAs and chitosan induced both new bone formation (**f**) and vascularization (**g**) compared with blank and pure chitosan [29]. *: $p < 0.05$, **: $p < 0.01$, ***: $p < 0.001$.

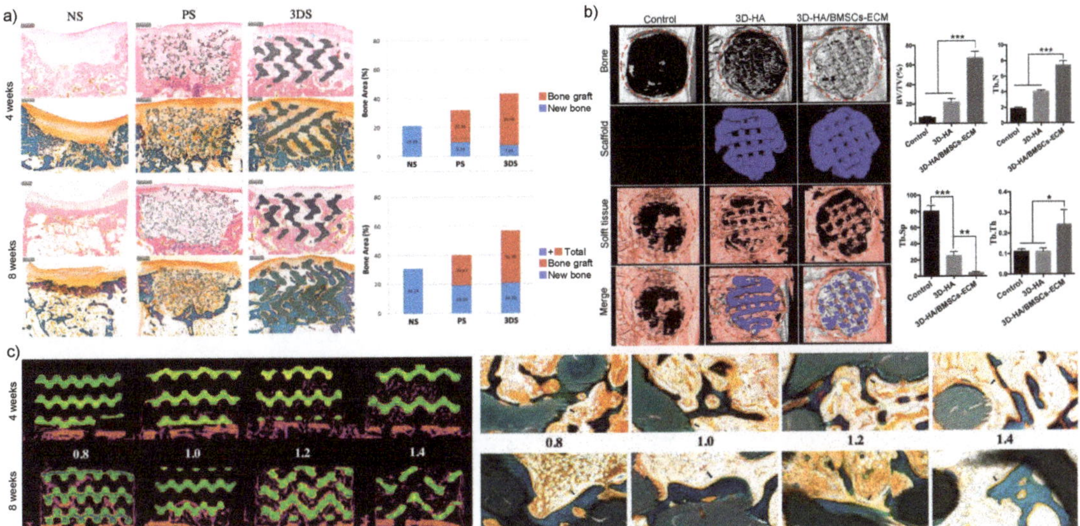

Figure 3. The properties affecting bone regeneration of nHA-based 3D printing scaffolds. (**a**) Histomorphometric examination demonstrated that the 3D printed HA/ tricalcium phosphate scaffold (3DS) had better bone reparation effect than both negative control (NS) and positive control (particle-type bone substitutes, PS) in beagle dog mandibular bone defects [30]. (**b**) Radiographic analysis showed that 3D hydroxyapatite/extracellular matrix (HA/ECM) composites promoted bone regeneration in rat calvarial defects after 12 weeks [34]. (**c**) Both radiographic and histological evaluations exhibited more new bone formation for 3D printed nHA-based scaffolds with higher pore size only at 4 weeks [35]. *: $p < 0.05$, **: $p < 0.01$, ***: $p < 0.001$.

3. Nano Silica

Nano silica is an important component of bioceramics, and it is widely applied in bone tissue engineering. According to the definition of the International Federation of Applied Chemistry, mesoporous materials have a pore size between 2 nm and 50 nm. Sol-gel method was initially used to synthesize regular pore structure in mesoporous silica nanoparticles (MSNs) with adjustable pore size [36]. The mesoporous structure offers good biocompatibility and biodegradability among inorganic nanomaterials [37]. Based on the exciting characteristics, MSNs are drawing more and more interest in the basic research of bone tissue engineering. MSNs serve as a kind of drug carrier with excellent release efficiency, and they also have the biological activity of promoting bone formation. Therefore, the researchers mainly focus on two aspects of modification: one is to modify the pore size adapted to functional factors that the mesoporous silicon material can load; the second is to explore different composite scaffold materials to improve the physiochemical property and bioactivity of MSNs.

The mechanisms of MSNs that promote bone defect reparation are as follows: (1) the silicon ions released by hydrolysis can promote the expression of osteogenic-related genes in osteoblasts (Figure 4a) [13]; (2) the mesoporous structure helps the deposition of HA, which further promotes mineralization [38]; (3) MSNs induce efficient macrophage uptake and promote immunomodulatory effects, which are conducive to osteogenic differentiation [39]. Moreover, the efficient uptake is good for drug delivery in bone defect area in vivo. The efficiency of drug delivery relies on the pore size and superficial chemical modification. As the controllability of MSN fabrication based on various synthesis methods has already been testified to load different types of drugs, the characteristics of modified MSN-drug releasing system are used more and more often in basic research with or without combined scaffolds.

MSN loaded-drugs including traditional chemicals, protein and peptides and nucleotides exert positive effects on bone formation. Combining dexamethasone-loaded MSNs with mineralized porous biocomposite scaffolds induced bone regeneration by enhancing the osteogenic activity of host BMSCs [40]. Alvarez et al. demonstrated that ibandronate sodium-loaded MSNs incorporated into the collagen gel had continuous drug release around 10 days, which effectively inhibited the function of osteoclasts and promoted the osteogenic differentiation of MSCs [41]. 17-beta estradiol-loaded MSNs with EDTA modified surface significantly enhanced the efficiency of hormone therapy for osteoporosis (Figure 4b) [42]. Furthermore, injectable miR222/MSN/aspirin hydrogel promoted mandibular bone regeneration and induced neurogenesis in the defect area [43]. Therefore, the effectiveness of drug delivery of MSNs is demonstrated by numerous studies and is a promising strategy for drug-MSN-scaffold-induced bone regeneration. To acquire maximum loading capacity of drugs, researchers modified the structure of MSNs with rough or hydrophilic surface, which enhanced the adhesion of the target drug [44]. Based on previous studies, hollow structure has enough space for drug loading and storage, thus hollow mesoporous silica materials provide new insight for mesoporous silica-based drug delivery [45]. However, the simple modification of surface or pore size of MSNs is not sufficient to realize precise releasing system. Therefore, the incorporation of biomacromolecules (organic polymers, macrocyclic molecules, etc.) and MSNs enhances the efficacy of MSN-drug releasing system.

3.1. Modification of MSN-Based Scaffolds

In order to acquire enough strength, proper morphology and enhanced bioactivity of MSNs for bone formation, the composites of MSNs and organic/inorganic scaffolds have been fabricated. Considering the importance of nHAs during biomineralization, numerous studies have adopted the combination of MSNs with nHAs and testified to the synergistic effects on bone formation. He et al. showed that MSN/nHA composites enhanced the adhesion and proliferation rate and osteogenesis-related gene expression of rabbit BMSCs and new bone formation in rabbit femur defects, compared to MSNs or nHAs alone [46]. Similar co-enhancement effect on osteogenic differentiation of MG-63 cells was found by Shuai et al. [47]. Furthermore, the existence of nHAs endowed discontinuous pore surface of MSNs, which induced rapid drug releasing of ciprofloxacin from 32% to 93% in 24 h [48]. Along with nHA-silica nanomaterials, the natural components of extracellular matrix can induce an increase of cell–scaffold interaction and result in better biocompatibility and bioactivity of nano silica. In turn, the application of silica reinforced the mechanical strength, enhanced the water uptake capacity, and fastened degradation rate of matrix scaffolds [49], which benefited repair of large-scale bone defects. Thus, the application of matrix components as silica-based scaffolds is attracting the interest of researchers. Gaihre et al. reported that injectable nano silica–chitosan microparticles performed significant enhancement of ALP activity during osteogenic induction of osteoblasts [50]. In addition, collagen-alginate-nano-silica microspheres improved the osteogenic potential of human osteoblast-like MG-63 cells by promoting the expression of OCN and BMP-2 [51]. Intrafibrillar silicified collagen scaffold promoted in situ bone regeneration via p38 signaling pathway in monocytes and recruited host MSCs (Figure 4c) [52]. Therefore, it is a promising method to fabricate nano silica with natural polymer scaffolds for bone regeneration applications.

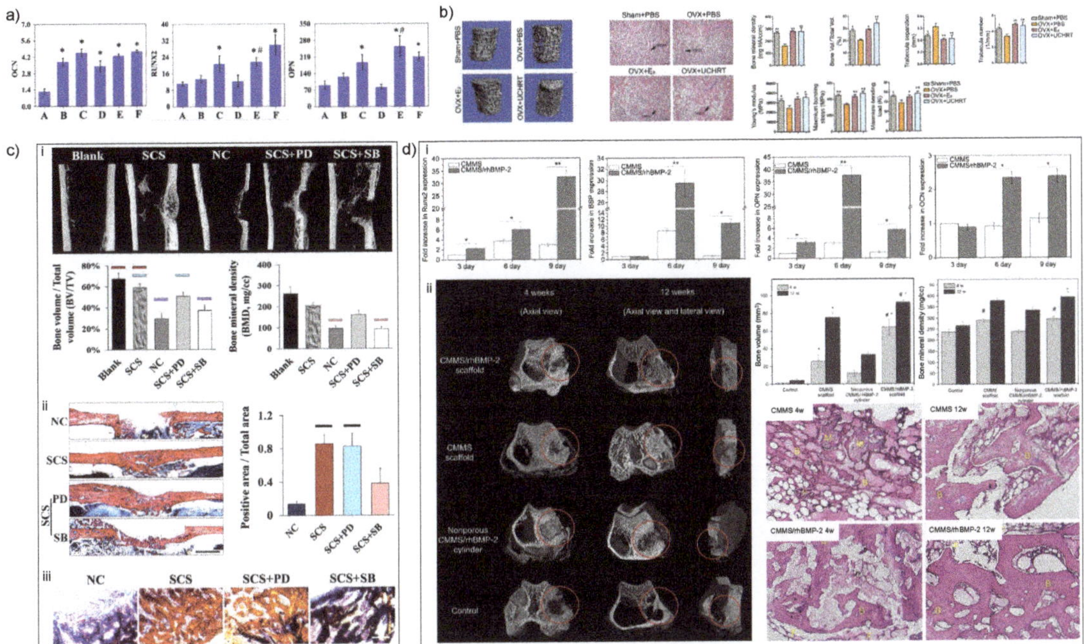

Figure 4. The application of nano silica in bone tissue engineering. (**a**) The effect of Si ions on osteogenesis-related gene expression of human bone marrow mesenchymal stem cells (BMSCs). A: blank control (DMEM); B: 10 µg/mL Si ions in DMEM; C: 50 µg/mL Si ions in DMEM; D–F: A–C solution plus 35 µg/mL dimethyloxaloylglycine, respectively. *: Comparison between blank and other groups; #: comparison between group B and E, C and F, respectively; $p < 0.05$ [13]. (**b**) Mesoporous silica nanoparticles (MSNs) assisted E2 sustained release, which prevented osteoporosis in vivo. OVX = ovariectomy; UCHRT = $NaLuF_4$:Yb,Tm@$NaLuF_4$@$mSiO_2$-EDTA-E2 nanocomposites. *: $p < 0.05$, **: $p < 0.01$ [42]. (**c**) Nano silica incorporated collagen fibrils promoted bone regeneration in rat femoral defects. SCS = silicified collagen scaffold; NC: negative control; SCS + PD: silicified collagen scaffold plus PD098059; SCS + SB: silicified collagen scaffold plus SB203580; PD and SB are MAPK inhibitors. (**i**) Micro-CT scans and quantitative results. (**ii**) van Geison staining. Bar = 1 mm. (**iii**) Ponceau trichrome staining. Bar = 200 µm [52]. (**d**) The osteoconductivity of calcium/magnesium-doped rhBMP-2-incorporated MSN scaffold in vitro and in vivo. CMMS= calcium/magnesium-doped silica-based scaffolds. (**i**) rhBMP-2 increased the osteogenesis-related gene expression of rat BMSCs. (**ii**) CMMS/rhBMP-2 scaffold promoted bone regeneration in rabbit femoral defects as demonstrated by micro-CT analysis and histological evaluation. M: materials, B: bone, F: fibrous tissue. *: $p < 0.05$; #: $p < 0.05$ [53].

Furthermore, some studies focus on doped metal ions in nano-silica-based materials. Dai et al. synthesized magnesium-doped mesoporous silica materials containing recombinant human bone morphogenetic protein-2 (rhBMP-2) with macroporous and mesoporous structure using polyurethane foam as a template (Figure 4d) [53]. This magnesium-doped silica material induced osteogenic differentiation of rat BMSCs in vitro and ectopic osteogenesis in vivo and also showed a good repair effect in a rabbit femoral defect model with diameter of 5 mm and depth of 5 mm. Shi et al. demonstrated that copper-doped MSNs induced robust immunomodulatory effects of murine-derived macrophage cell line RAW 264.7 and promoted osteogenesis of human BMSCs [39]. Recently, rare earth elements, such as lanthanum- [54], europium- [55] and gadolinium-doped [56] MSNs showed the exciting ability in osteogenesis of BMSCs and bone regeneration in rat skull bone defects. Despite the positive results in osteogenesis in vitro, the application of metal ions in nano silica materials still need more in vivo evidence to demonstrate the effectiveness and biosafety of such modification of nano silica.

4. Metallic Nanomaterials

Most traditional bulk metals exhibit much higher mechanical properties and worse bioactivity than natural bone, which may lead to bone resorption and poor osteointegration and osteogenesis [9]. Surface modification at nanoscale would help to improve the surface topography and chemistry of metal implants. In order to balance the gap in mechanical properties, the researchers tried to build a micro-nano structure to increase the porosity of metal materials to more than 50% [57]. In the meanwhile, porous materials can promote the penetration of cells and nutrients and the regeneration of bone and blood vessel when the pore size exceeds 300 μm [58]. In addition, the metallic nanomaterials are preferably biodegradable, which means they are expected to gradually corrode in vivo, and the corrosion products are metabolized or absorbed by cells and/or tissues. After the defects are repaired, the implants are completely dissolved without residue. In the following sections, four widely studied metal nanomaterials are introduced.

4.1. Ti-Based Nanomaterials

4.1.1. Nanoscale Surface Modification of Ti-Based Biomaterials

Titanium (Ti)-based biomaterials are widely used as permanent implants in orthopedic surgery and dental implantation because of their high load-bearing properties, nondegradability and good biocompatibility [59]. Due to the superior corrosion resistance of bulk Ti and its alloy, they often exhibit slow biological response, low osseointegration rate and lack of antibacterial properties. Nanoscale surface modification strategies, such as coating and doping, are effective means to solve the above problems and endow the implants with functionalization. In order to enhance the osteoconductivity, Ding et al. mixed strontium-incorporated lysozyme solution with Ti substrates and spontaneously formed a 2D nanofilm which could promote the ALP activity and osteogenesis-related genes expression of BMSCs [60]. The sustained release of Sr^{2+} from lysozyme nanofilm further facilitated osteointegration of Ti implants in rat femur bone defect model. Similarly, nano-graphene oxide (GO) was deposited on Ti surface through ultrasonic atomization spraying technique. This nano-GO coating induced the osteogenic differentiation of rat BMSCs via FAK/P38 signaling pathways and further accelerated bone regeneration in vivo [61]. Furthermore, lanthanide mineral-substituted hydroxyapatite nanorods were coated on Ti substrates by electrophoretic deposition method to mimic the topography and composition of natural bone [62]. The lanthanide mineral-substituted hydroxyapatite nanorods modified Ti implants had better osteogenic effect in vitro and bone regeneration effect in rat tibia defects. Compared to the above-mentioned inorganic coatings, human MSC-derived extracellular vesicles (MSC-EVs) were difficult to be anchored onto pristine Ti. Chen et al. successfully self-assembled human MSC-EVs onto biotin-doped polypyrrole titanium, which exhibited osteoinductivity in nude mice ectopic bone formation model with the help of osteoinductive miRNAs tested in MSC-EVs [63]. As for the antimicrobial activity, Zhang et al., inspired by the adhesion mechanism of mussel, developed a novel silicon-doped calcium phosphate composite coating (Van-pBNPs/pep@pSiCaP) loaded with vancomycin on porous Ti scaffold via modified surface mineralization process [64]. The functionalized biomimetic Ti scaffold can prevent the adhesion and proliferation of *staphylococcus epidermidis*. In addition to antibiotics, metal ions also have bactericidal activity. Huang et al. fabricated Cu-containing micro/nanotopographical bioceramic surface through micro-arc oxidation. The subsequent hydrothermal and final heat treatment can assist Ti implants to induce proinflammatory M1 macrophage polarization via Cu-transport signaling pathway and enhance bacterial phagocytosis. [65]. In addition to chemical antibacterial, the researchers prepared zinc oxide@collagen type I coating to achieve a broad-spectrum antibacterial effect through the photothermal effect of zinc oxide [66]. Surface modification materials can also cooperate with metal matrix to promote bone tissue repair. Fu et al. proved that the silicon-doped hydroxyapatite coating could demonstrate not only enhanced osteogenesis, but also promote angiogenesis [67].

4.1.2. Additive Manufacturing of Ti-Based Biomaterials

Traditional bulk Ti implants have the advantages of easy commercial availability and wide clinical acceptance, but their much higher stiffness than natural bone may cause stress shielding, leading to bone resorption and implant failure [68]. Although investigators are committed to improving the surface topography and chemical structure of bulk Ti implants, the porous structure, essential to the mechanical properties of natural bone, could not be simulated through traditional manufacture methods. The emergence of AM technologies is expected to solve this problem fundamentally. The computer-controlled and bottom-up fabrication process of AM technologies can more easily realize the control of chemical composition and micro-nano structure, so as to obtain better biomechanical properties to promote bone regeneration [69]. The elastic modulus of Ti–6Al–4V–10Mo [70], Ti–35Nb [71] and Ti–50Ta [72] (in wt%) nanomaterials prepared by 3D printing was significantly reduced to 73–84.7 GPa. The 3D printed, customized Ti implants have been used in craniofacial and orthopedic applications [73]. A clinical study involving 21 patients demonstrated good fixation between bone and custom-made 3D-printed Ti implants (with surface area ranging from 12,146 to 24,980 mm^2) and satisfactory skull-shape symmetry without any complications during the 6 to 24 months follow-up period (Figure 5a) [74]. For tongue cancer excision with a tumor recurrence, Lee et al. utilized Ti-6AL-4V-ELI medical grade powder to fabricate a Ti implant with four dental implants based on computer-aided design data and restored the facial symmetry and occlusion at 5 months after surgery [75]. Impressively, Ti-6AL-4V-based AM implant with a total volume of 30.7 cm^3 was able to successfully replace the complicated bone defects of distal tibia and foot caused by motor vehicle collision, and the patients could basically return to normal life and walk without ambulatory aids after 6 months [76]. Although there are still early and late complications, cranioplasty based on AM Ti implants has a higher success rate compared with other techniques [77]. In order to modify porous Ti nanomaterials based on AM, geometrical cues and surface structure design were investigated. As the porosity of AM Ti implants increased from 55.51% to 76.14%, the yield compressive strength decreased from 186.05 ± 1.85 MPa to 51.95 ± 0.62 MPa, and the elastic moduli decreased from 6.74 ± 0.47 GPa to 0.98 ± 0.03 GPa [78]. Nonetheless, the sharp decline of mechanical strength did not affect the bone regeneration, which may be attributed to the optimum porosity. Interestingly, the regeneration results of implanted non-load-bearing bones (such as the skull) were much worse than those of load-bearing bones such as the femur, which indicated the importance of stress stimulation [78]. In contrast, Maietta et al. demonstrated that with the help of finite element analysis, the Ti-6AL-4V architectural characteristics, such as pore size and shape, could be changed without significant alteration in mechanical properties [79]. For biological applications, Zhu et al. proved that shape- and size-controlled microgroove-patterned Ti surface structure manufactured by a combination of photolithography and inductively coupled plasma-based dry etching was beneficial to osteogenesis and bone regeneration [80]. Specifically, R3G7 (ridge width of 3 µm, groove width of 7 µm and depth of 2 µm) was the most effective micropattern to promote the osteogenic differentiation of MC3T3-E1 cells and bone regeneration in rat calvarial defects (Figure 5d). In addition to porosity and micropattern, roughness also plays an important role. Saruta et al. found that submicro-rough surface (with roughness of 24 ± 1.2 nm) promoted the attachment, proliferation and calcium deposition of osteoblasts, while micro-rough surface (with roughness of 123 ± 6.15 nm) had the strongest bone–implant integration in rat femurs [81]. Surface biofunctionalization is also well investigated to modify AM porous Ti-based nanomaterials. Several kinds of coating including silk fibroin, calcium phosphate and tricalcium phosphate can promote bone integration and regeneration both in normal [64,82] and osteoporotic bone defect models [83] (Figure 5b,c). The vancomycin coating helps to prevent the bacteria colonization caused by the high surface area to volume ratio of porous Ti implants [82]. On the other hand, a high surface area/volume ratio is beneficial for drug sustained release, such as BMP-2 and silver nanoparticles [84].

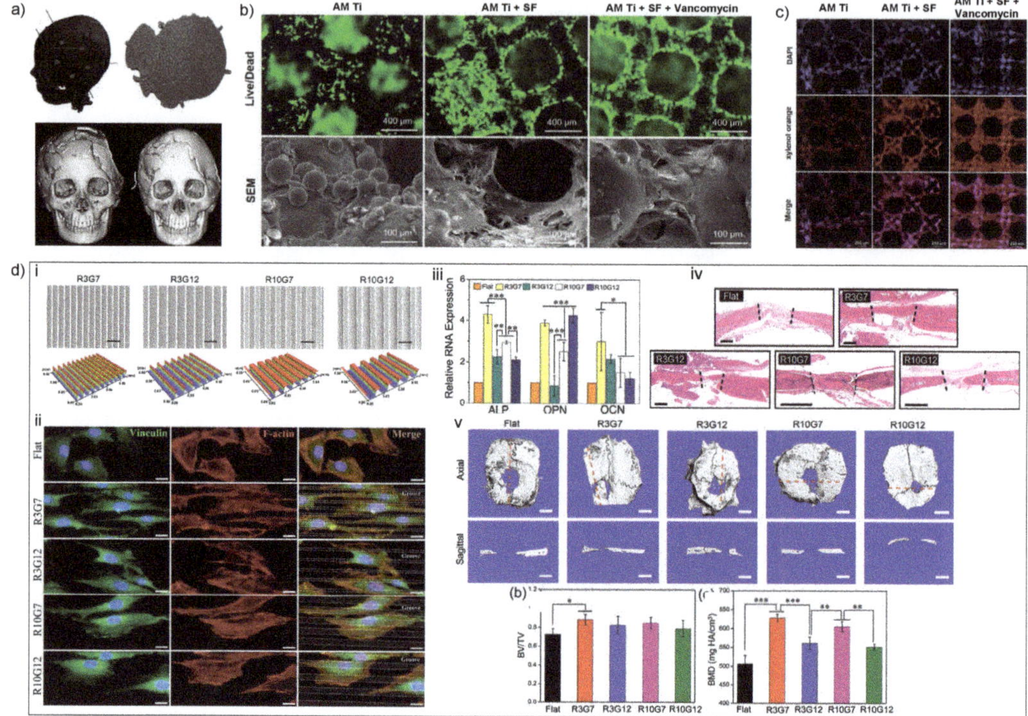

Figure 5. The biological properties of AM Ti-based implants. (**a**) Custom-made Ti-6Al-4V-ELI AM implants helped to reconstruct the skull defect of 21 patients [74]. (**b**) Silk fibroin and vancomycin coating promoted the survival of MC3T3-E1 cell line seeded on 3D printed Ti implants [82].(**c**) Silk fibroin and vancomycin coating promoted the calcium deposition of MC3T3-E1 cell line seeded on 3D printed Ti implants [82]. (**d**) The micropattern of Ti substrates induced osteogenesis. (**i**) SEM images and 3D surface profile of different micropatterns. (**ii**) MC3T3-E1 preosteoblasts aligned along the ridges in R3G7 micropattern. (**iii**) The R3G7 microgroove pattern promoted osteogenesis of MC3T3-E1 cells. (**iv**) and (**v**) The R3G7 microgroove pattern induced bone regeneration in rat skull defects [80]. *: $p < 0.05$; **: $p < 0.01$; ***: $p < 0.001$.

As mentioned above, AM Ti-based nanomaterials would have broad application aspects in bone tissue regeneration. The excellent chemical stability of titanium makes it difficult to degrade in vivo. Although titanium-based biomaterials are still the mainstream of current clinical applications, the developments of biodegradable metallic nanomaterials are promising and necessary.

4.2. Mg-Based Biomaterials

4.2.1. Nanoscale Surface Modification of Mg-Based Biomaterials

Magnesium is the most studied biodegradable metal ion. It has been confirmed that magnesium ions can induce bone marrow MSCs to differentiate into osteoblast lineage through canonical Wnt signaling pathway [85]. In distraction osteogenesis model, high-purity magnesium pins promoted angiogenesis and bone consolidation [86]. Moreover, the elastic modulus of magnesium is much lower than that of Ti, preventing bone resorption and implant failures induced by stress shielding. Additionally, the excellent degradability of Mg makes it an excellent temporary bone fixation device and possible low load-bearing bone substitute [87]. Degradability is a double-edged sword. In most cases, pure Mg implants degrade too fast to fully support new bone formation and the rapid release of hydrogen may interfere with the local microenvironment and hinder bone regeneration. Therefore, alloys of magnesium with calcium [7], zinc [7], strontium [88] and rare earth (RE) elements [89] have been developed to solve the above-mentioned problems. A long-term clinical study has confirmed the early bone healing and the complete replacement of Mg-5wt%Ca-1wt%Zn alloy by new bone in the late stage [7]. Fifty-three patients with hand and wrist fractures were involved in this study. All the patients returned to normal life with no sign of pain and the Mg-5wt%Ca-1wt%Zn implants were degraded, as confirmed by radiographic examination within 1 year, demonstrating the controlled degradation of this kind of Mg-based alloy. Furthermore, in order to reduce the corrosion rate of Mg, various nanoscale surface coating strategies have been applied to provide Mg-based alloys with a protective layer [90]. Zhang et al. developed calcium-phosphate-coated Mg−Zn−Gd scaffolds via chemical deposition method, which could not only repair rat cranial defects of critical size, but also promote osteogenesis, angiogenesis and the production of neuropeptide calcitonin gene-related peptide from trigeminal neurons in the orbital bone defect model of beagle dog [91]. Similarly, polycaprolactone and nHA nanocomposites dual coating via dip-coating and electrospinning could induce bone regeneration in rabbit femoral defects [92]. Kang et al. fabricated poly(ether imide)–SiO_2/nHA-coated porous Mg scaffold through dip-coating technique, and repaired the bone defect in the femoropatellar groove model of rabbits due to the proper corrosion rate and enhanced mechanical strength of hybrid scaffold [12]. In addition, the surface modification by inorganic nanomaterials can also bring new functions to the implanted scaffold. For example, a functionalized 70-nm-thick TiO_2/Mg_2TiO_4 nanolayer was fabricated by plasma ion immersion implantation technique on WE43 Mg-based alloy. The TiO_2/Mg_2TiO_4 coating can promote osteogenesis of MC3T3-E1 cells and induce twice and six times higher levels of new bone volume (175%) than those of pristine WE43 alloy (88%) and blank control (28%) in rat femoral defect model. In the meanwhile, TiO_2/Mg_2TiO_4 nanolayer suppressed bacterial infection and controlled the degradation behavior [93]. As for the immumodulatory effect, the combination of fibrinogen and magnesium can lead macrophages to M2 polarization and further promote osteogenic differentiation of MSCs [94]. In some special bone defect models (such as osteoporotic fracture), biodegradable Mg-based implants may serve as a sustained drug delivery system. The calcium phosphate nanocoating ensured the suitable degradation rate and structural integrity of Mg-based alloy, and the co-delivered magnesium degradation products and zoledronic acid modulated bone formation and resorption [95].

4.2.2. Additive Manufacturing of Mg-Based Biomaterials

Compared to bulk Mg-based biomaterials, the AM of biodegradable Mg implants is still quite difficult due to the intrinsic properties of magnesium. Salehi et al. developed a novel 3D printing technique followed by sintering process to fabricate the Mg-5.9Zn-0.13Zr alloy with high precision at nano- and micron-scale. This AM Mg-based alloy possessed an average pore diameter of 15 μm, compressive strength of 174 MPa and elastic modulus of 18 GPa, which were quite similar to those of human cortical bone [96]. Until now, only a few research groups fabricated Mg alloys through selective laser melting technique [97] and investigated their mechanical and degradation properties. The biological performances of AM Mg-based alloys, for instance biocompatibility and regenerative effect in vivo, need to be further investigated.

4.3. Zn-Based Biomaterials

Zinc is an essential element for humans and plays an important role in many physiological activities. Zinc ions can enter human MSCs with the help of TRPM7 and GPR39, and then activate cAMP-PKA pathway, thereby ultimately enhancing cell survival/proliferation, differentiation, ECM mineralization and osteogenesis [98]. The degradation rate of Zn-based biomaterials is slower than that of Mg, and matches the tissue healing speed. The mechanical properties of Zn materials are between Ti and Mg, similar to natural bone [99]. Biodegradable Zn alloys, such as Zn-0.8%Li- (Mg, Ag), Zn–2Ag–1.8Au–0.2V and Zn-1Ag have better mechanical properties, corrosion rates and antibacterial activities than pure Zn implants, while only certain alloys have modest biocompatibility [100]. After the first published work in 2017 [101], a growing number of AM Zn-based nanomaterials appeared. The mechanical properties of AM porous Zn are similar to those of cancellous bone, and the degradation rates allow satisfactory bone substitution [102]. Although Zn exhibits mild cytotoxicity, it is still acceptable [103]. Like Mg, the biological evaluation of Zn-based biomaterials, especially AM porous Zn and alloys, remains blank.

4.4. Au-Based Biomaterials

Gold nanoparticles (AuNPs) have been widely investigated in biomedical applications, especially in antitumor therapy and bone regeneration therapy, due to their good biocompatibility, photothermal stability and near-infrared absorbance. The size and morphology of gold nanomaterials are important factors affecting their biological functions. Celentano et al. have successfully synthesized stable ultra-small gold nanoparticles, anisotropic gold nanoflowers and twisted gold nanorods by a simple green method and confirmed the biocompatibility of these AuNPs [104–106]. The addition of AuNPs into poly(methyl methacrylate)-based bone cement can significantly improve the punching performances while maintaining stable compressive properties [107]. In addition to better mechanical properties, as an efficient photothermal agent, AuNPs can eradicate residual tumor cells after the solid tumor is removed through photothermal therapy [108]. The theory of photothermal therapy can also be used to treat osteomyelitis. Gold-nanocage-containing aspirin can convert laser light into heat and realize controlled release of aspirin, which can perform anti-inflammatory effects on monocytes and promote bone regeneration after the monocytes eliminate infection [109]. Similarly, Sanchez-Casanova et al. entrapped heat-activated transgenic cell constructs in near-infrared-responsive hydrogels containing AuNPs, which can conditionally produce BMP-2 and promote bone regeneration in vivo [110]. Despite the above progress, it is still necessary to clarify how to co-assemble AuNPs and biomolecules and form biomimetic hierarchical scaffolds to achieve better outcomes in the field of bone regeneration therapy.

5. Concluding Remarks and Future Perspectives

Trauma, infection, tumor, degenerative and congenital diseases are the major causes of excessive bone defects. The current bone tissue engineering is trying to develop new materials to replace autologous bone grafts to achieve bone regeneration. Novel therapies based on inorganic nanomaterials, providing excellent mechanical properties and abundant physical/chemical/biological functions, are complementary to natural biomacromolecules and polymeric-based materials (Table 1). The development of inorganic nanoparticle/polymer composites can effectively integrate the respective advantages of inorganic and organic phases, bringing unlimited possibilities for the development of novel bone substitute materials. Biomimetic mineralization utilizing nano hydroxyapatites and nano silica is currently one of the most successful organic material modification processes. Furthermore, with the progress of technology, precisely designed porous inorganic nanomaterials appear to have lower stiffness and elastic modulus than traditional bulk materials, making them much closer to natural bone. Additive manufacturing is the most promising technology to realize computer-aided design and personalized customization. The 3D printed nHA-based nanomaterials and Ti-, Mg-, Zn-based metals would be able to act as temporary fixations or permanent implants according to their own characteristics, and eventually be replaced by regenerated bone tissue. Finally, inorganic nanoparticles are excellent carriers of drugs, growth factors and genes. nHAs, MSNs and AuNPs are commonly used to carry drugs into scaffolds and realize sustained release with the controllable degradation of scaffolds.

Table 1. The advantages and drawbacks of novel inorganic nanomaterial.

Type	Advantages		Drawbacks
nHA/polymer composites	(1) (2) (3)	Enhanced mechanical performance of polymers [16,19–23] Delayed degradation rate of polymers [25,26] Enhanced osteogenesis [24,27–29]	Not custom-made
3D printed nHA-based inorganic nanomaterials	(1) (2)	Customized [30] Flexible mechanical and biological properties [31,32]	Immature design and manufacturing methods [35]
MSNs	(1) (2) (3) (4)	Good biocompatibility [37] Good biodegradability [37] Sustained release of silicon ions [13], drugs [40–42], cytokines and miRNAs [43] Immunomodulatory effects [39]	Seldom used alone in bone regeneration
Ti-based nanomaterials	(1) (2) (3)	High load-bearing properties [59] Good biocompatibility [59] AM Ti nanomaterials could change the exorbitant elastic modulus of traditional Ti materials [69–72] and replace complicated bone defects [74–76].	(1) Nondegradability [59] (2) Poor biological response and anti-bacterial properties [59]
Mg-based biomaterials	(1) (2)	Biodegradability [86] Osteoinductivity [85,86]	(1) Fast degradation and hydrogen release rate (2) Low elastic modulus for load-bearing bone defects [87] (3) Immature AM technology of Mg-based biomaterials
Zn-based biomaterials	(1) (2)	Biodegradability and suitable degradation rate Suitable mechanical properties similar to natural bone [99]	(1) Mild cytotoxicity [103] (2) Immature AM technology of Zn-based biomaterials
AuNPs	(1) (2) (3)	Good biocompatibility [108] Photothermal stability [108] Near-infrared absorbance [108]	(1) Seldom used alone in bone regeneration (2) The co-assembly of AuNPs and biomolecules lacks hierarchical structure

Author Contributions: D.L. and Y.L. designed, prepared and revised the manuscript; Y.F. and S.C. wrote the manuscript. All authors have read and agreed to the published version of the manuscript.

Funding: The authors are grateful for financial support from the Projects of Ten-Thousand Talents Program No. QNBJ2019-2 (Y.L.), Key R&D plan of Ningxia Hui Autonomous Region No. 2020BCG01001 (Y.L.), National Natural Science Foundations of China No. 81871492 (Y.L.), No. 51902344 (D.L.) and No. 81901053 (Y.F.), the National Key R&D project from Minister of Science and Technology, China No 2016YFA0202703 (D.L.).

Conflicts of Interest: The authors declare no conflict of interest.

References

1. Wegst, U.G.; Bai, H.; Saiz, E.; Tomsia, A.P.; Ritchie, R.O. Bioinspired structural materials. *Nat. Mater.* **2015**, *14*, 23–36. [CrossRef] [PubMed]
2. Roddy, E.; DeBaun, M.R.; Daoud-Gray, A.; Yang, Y.P.; Gardner, M.J. Treatment of critical-sized bone defects: Clinical and tissue engineering perspectives. *Eur. J. Orthop. Surg. Traumatol.* **2018**, *28*, 351–362. [CrossRef]
3. Griffin, K.S.; Davis, K.M.; McKinley, T.O.; Anglen, J.O.; Chu, T.-M.G.; Boerckel, J.D.; Kacena, M.A. Evolution of Bone Grafting: Bone Grafts and Tissue Engineering Strategies for Vascularized Bone Regeneration. *Clin. Rev. Bone Miner. Metab.* **2015**, *13*, 232–244. [CrossRef]
4. Burk, T.; Del Valle, J.; Finn, R.A.; Phillips, C. Maximum Quantity of Bone Available for Harvest From the Anterior Iliac Crest, Posterior Iliac Crest, and Proximal Tibia Using a Standardized Surgical Approach: A Cadaveric Study. *J. Oral Maxillofac. Surg.* **2016**, *74*, 2532–2548. [CrossRef]

5. Cross, L.M.; Thakur, A.; Jalili, N.A.; Detamore, M.; Gaharwar, A.K. Nanoengineered biomaterials for repair and regeneration of orthopedic tissue interfaces. *Acta Biomater.* **2016**, *42*, 2–17. [CrossRef]

6. Morse, D.E. Silicon biotechnology: Harnessing biological silica production to construct new materials. *Trends Biotechnol.* **1999**, *17*, 230–232. [CrossRef]

7. Lee, J.W.; Han, H.S.; Han, K.J.; Park, J.; Jeon, H.; Ok, M.R.; Seok, H.K.; Ahn, J.P.; Lee, K.E.; Lee, D.H. Long-term clinical study and multiscale analysis of in vivo biodegradation mechanism of Mg alloy. *Proc. Natl. Acad. Sci. USA* **2016**, 716. [CrossRef] [PubMed]

8. Ljungblad, U. Statistical process control applied to additive manufacturing enables series production of orthopedic implants. In Proceedings of the 21st International DAAAM Symposium, Zadar, Croatia, 20–23 October 2010; DAAAM International Vienna: Vienna, Austria, 2010.

9. Roseti, L.; Parisi, V.; Petretta, M.; Cavallo, C.; Desando, G.; Bartolotti, I.; Grigolo, B. Scaffolds for Bone Tissue Engineering: State of the art and new perspectives. *Mat. Sci Eng. C Mater.* **2017**, *78*, 1246–1262. [CrossRef]

10. Liu, Y.; Luo, D.; Yu, M.; Wang, Y.; Jin, S.; Li, Z.; Cui, S.; He, D.; Zhang, T.; Wang, T.; et al. Thermodynamically Controlled Self-Assembly of Hierarchically Staggered Architecture as an Osteoinductive Alternative to Bone Autografts. *Adv. Funct. Mater.* **2019**, *29*, 1806445. [CrossRef]

11. Li, Y.; Pavanram, P.; Zhou, J.; Lietaert, K.; Taheri, P.; Li, W.; San, H.; Leeflang, M.A.; Mol, J.M.C.; Jahr, H.; et al. Additively manufactured biodegradable porous zinc. *Acta Biomater.* **2020**, *101*, 609–623. [CrossRef] [PubMed]

12. Kang, M.H.; Lee, H.; Jang, T.S.; Seong, Y.J.; Kim, H.E.; Koh, Y.H.; Song, J.; Jung, H.D. Biomimetic porous Mg with tunable mechanical properties and biodegradation rates for bone regeneration. *Acta Biomater.* **2019**, *84*, 453–467. [CrossRef] [PubMed]

13. Shi, M.; Zhou, Y.; Shao, J.; Chen, Z.; Song, B.; Chang, J.; Wu, C.; Xiao, Y. Stimulation of osteogenesis and angiogenesis of hBMSCs by delivering Si ions and functional drug from mesoporous silica nanospheres. *Acta Biomater.* **2015**, *21*, 178–189. [CrossRef]

14. Qiao, S.; Wu, D.; Li, Z.; Zhu, Y.; Zhan, F.; Lai, H.; Gu, Y. The combination of multi-functional ingredients-loaded hydrogels and three-dimensional printed porous titanium alloys for infective bone defect treatment. *J. Tissue Eng.* **2020**, *11*, 2041731420965797. [CrossRef] [PubMed]

15. Koons, G.L.; Diba, M.; Mikos, A.G. Materials design for bone-tissue engineering. *Nat. Rev. Mater.* **2020**. [CrossRef]

16. Liu, Y.; Luo, D.; Kou, X.I.; Wang, X.; Zhou, Y. Hierarchical Intrafibrillar Nanocarbonated Apatite Assembly Improves the Nanomechanics and Cytocompatibility of Mineralized Collagen. *Adv. Funct. Mater.* **2012**, *23*, 1404–1411. [CrossRef]

17. Fu, Y.; Liu, S.; Cui, S.J.; Kou, X.X.; Wang, X.D.; Liu, X.M.; Sun, Y.; Wang, G.N.; Liu, Y.; Zhou, Y.H. The surface chemistry of nanoscale mineralized collagen regulates periodontal ligament stem cell fate. *ACS Appl. Mater. Interfaces* **2016**, 15958. [CrossRef] [PubMed]

18. Liu, Y.; Liu, S.; Luo, D.; Xue, Z.; Yang, X.; Gu, L.; Zhou, Y.; Wang, T. Hierarchically Staggered Nanostructure of Mineralized Collagen as a Bone-Grafting Scaffold. *Adv. Mater.* **2016**, *28*, 8740–8748. [CrossRef]

19. Sionkowska, A.; Kaczmarek, B. Preparation and characterization of composites based on the blends of collagen, chitosan and hyaluronic acid with nano-hydroxyapatite. *Int. J. Biol. Macromol.* **2017**, *102*, 658–666. [CrossRef]

20. Nazeer, M.A.; Yilgor, E.; Yilgor, I. Intercalated chitosan/hydroxyapatite nanocomposites: Promising materials for bone tissue engineering applications. *Carbohydr. Polym.* **2017**, *175*, 38–46. [CrossRef]

21. Uswatta, S.P.; Okeke, I.U.; Jayasuriya, A.C. Injectable porous nano-hydroxyapatite/chitosan/tripolyphosphate scaffolds with improved compressive strength for bone regeneration. *Mater. Sci Eng. C* **2016**, *69*, 505–512. [CrossRef]

22. Liu, H.; Xu, G.W.; Wang, Y.F.; Zhao, H.S.; Xiong, S.; Wu, Y.; Heng, B.C.; An, C.R.; Zhu, G.H.; Xie, D.H. Composite scaffolds of nano-hydroxyapatite and silk fibroin enhance mesenchymal stem cell-based bone regeneration via the interleukin 1 alpha autocrine/paracrine signaling loop. *Biomaterials* **2015**, *49*, 103–112. [CrossRef]

23. Moeini, S.; Mohammadi, M.R.; Simchi, A. In-situ solvothermal processing of polycaprolactone/hydroxyapatite nanocomposites with enhanced mechanical and biological performance for bone tissue engineering. *Bioact. Mater.* **2017**, *2*, 146–155. [CrossRef] [PubMed]

24. Yu, J.; Xu, Y.; Li, S.; Seifert, G.V.; Becker, M.L. Three-Dimensional Printing of Nano Hydroxyapatite/Poly(ester urea) Composite Scaffolds with Enhanced Bioactivity. *Biomacromolecules* **2017**, *18*, 4171–4183. [CrossRef] [PubMed]

25. Higuchi, J.; Fortunato, G.; Wozniak, B.; Chodara, A.; Domaschke, S.; Meczynska-Wielgosz, S.; Kruszewski, M.; Dommann, A.; Lojkowski, W. Polymer Membranes Sonocoated and Electrosprayed with Nano-Hydroxyapatite for Periodontal Tissues Regeneration. *Nanomaterials* **2019**, *9*, 1625. [CrossRef] [PubMed]

26. Kumar, P.; Saini, M.; Dehiya, B.S.; Umar, A.; Sindhu, A.; Mohammed, H.; Al-Hadeethi, Y.; Guo, Z. Fabrication and in-vitro biocompatibility of freeze-dried CTS-nHA and CTS-nBG scaffolds for bone regeneration applications. *Int. J. Biol. Macromol.* **2020**, *149*, 1–10. [CrossRef]

27. Bohner, M.; Miron, R.J. A proposed mechanism for material-induced heterotopic ossification. *Mater. Today* **2019**. [CrossRef]

28. Gonzalez Ocampo, J.I.; Machado de Paula, M.M.; Bassous, N.J.; Lobo, A.O.; Ossa Orozco, C.P.; Webster, T.J. Osteoblast responses to injectable bone substitutes of kappa-carrageenan and nano hydroxyapatite. *Acta Biomater.* **2019**, *83*, 425–434. [CrossRef]

29. Feng, C.; Xue, J.; Yu, X.; Zhai, D.; Lin, R.; Zhang, M.; Xia, L.; Wang, X.; Yao, Q.; Chang, J.; et al. Co-inspired hydroxyapatite-based scaffolds for vascularized bone regeneration. *Acta Biomater.* **2020**. [CrossRef]

30. Kim, J.W.; Yang, B.E.; Hong, S.J.; Choi, H.G.; Byeon, S.J.; Lim, H.K.; Chung, S.M.; Lee, J.H.; Byun, S.H. Bone Regeneration Capability of 3D Printed Ceramic Scaffolds. *Int. J. Mol. Sci.* **2020**, *21*, 4837. [CrossRef]

31. Kim, Y.; Lee, E.J.; Davydov, A.V.; Frukhbeyen, S.; Seppala, J.E.; Takagi, S.; Chow, L.; Alimperti, S. Biofabrication of 3D printed hydroxyapatite composite scaffolds for bone regeneration. *Biomed. Mater.* **2020**. [CrossRef]

32. Wei, Y.; Liu, L.; Gao, H.; Shi, X.; Wang, Y. In Situ Formation of Hexagon-like Column Array Hydroxyapatite on 3D-Plotted Hydroxyapatite Scaffolds by Hydrothermal Method and Its Effect on Osteogenic Differentiation. *ACS Appl. Bio Mater.* **2020**, *3*, 1753–1760. [CrossRef]

33. Wei, Y.; Gao, H.; Hao, L.; Shi, X.; Wang, Y. Constructing a Sr(2+)-Substituted Surface Hydroxyapatite Hexagon-Like Microarray on 3D-Plotted Hydroxyapatite Scaffold to Regulate Osteogenic Differentiation. *Nanomaterials* **2020**, *10*, 1672. [CrossRef]

34. Chi, H.; Chen, G.; He, Y.; Chen, G.; Tu, H.; Liu, X.; Yan, J.; Wang, X. 3D-HA Scaffold Functionalized by Extracellular Matrix of Stem Cells Promotes Bone Repair. *Int. J. Nanomed.* **2020**, *15*, 5825–5838. [CrossRef]

35. Lim, H.K.; Hong, S.J.; Byeon, S.J.; Chung, S.M.; On, S.W.; Yang, B.E.; Lee, J.H.; Byun, S.H. 3D-Printed Ceramic Bone Scaffolds with Variable Pore Architectures. *Int. J. Mol. Sci.* **2020**, *21*, 6942. [CrossRef]

36. Kresge, C.T.; Leonowicz, M.E.; Roth, W.J.; Vartuli, J.C.; Beck, J.S. Ordered mesoporous molecular sieves synthesized by a liquid-crystal template mechanism. *Nature* **1992**. [CrossRef]

37. Meng, H.; Xue, M.; Xia, T.; Ji, Z.; Tarn, D.Y.; Zink, J.I.; Nel, A.E. Use of size and a copolymer design feature to improve the biodistribution and the enhanced permeability and retention effect of doxorubicin-loaded mesoporous silica nanoparticles in a murine xenograft tumor model. *Acs Nano* **2011**, *5*, 4131–4144. [CrossRef]

38. Shadjou, N.; Hasanzadeh, M. Bone tissue engineering using silica-based mesoporous nanobiomaterials: Recent progress. *Mater. Sci. Eng. C* **2015**, *55*, 401–409. [CrossRef] [PubMed]

39. Shi, M.; Chen, Z.; Farnaghi, S.; Friis, T.; Mao, X.; Xiao, Y.; Wu, C. Copper-doped mesoporous silica nanospheres, a promising immunomodulatory agent for inducing osteogenesis. *Acta Biomater.* **2016**, *30*, 334–344. [CrossRef] [PubMed]

40. Zhou, X.; Liu, P.; Nie, W.; Peng, C.; Wang, J. Incorporation of dexamethasone-loaded mesoporous silica nanoparticles into mineralized porous biocomposite scaffolds for improving osteogenic activity. *Int. J. Biol. Macromol.* **2020**, *149*, 116–126. [CrossRef] [PubMed]

41. Alvarez, G.S.; Alvarez Echazu, M.I.; Olivetti, C.E.; Desimone, M.F. Synthesis and Characterization of Ibandronate-Loaded Silica Nanoparticles and Collagen Nanocomposites. *Curr. Pharm. Biotechnol.* **2015**, *16*, 661–667. [CrossRef]

42. Chen, X.; Zhu, X.; Hu, Y.; Yuan, W.; Qiu, X.; Jiang, T.; Xia, C.; Xiong, L.; Li, F.; Gao, Y. EDTA-Modified 17beta-Estradiol-Laden Upconversion Nanocomposite for Bone-Targeted Hormone Replacement Therapy for Osteoporosis. *Theranostics* **2020**, *10*, 3281–3292. [CrossRef]

43. Lei, L.; Liu, Z.; Yuan, P.; Jin, R.; Wang, X.; Jiang, T.; Chen, X. Injectable colloidal hydrogel with mesoporous silica nanoparticles for sustained co-release of microRNA-222 and aspirin to achieve innervated bone regeneration in rat mandibular defects. *J. Mater. Chem. B* **2019**. [CrossRef]

44. Amoozgar, Z.; Yeo, Y. Recent advances in stealth coating of nanoparticle drug delivery systems. *Wiley Interdiscip. Rev. Nanomed. Nanobiotechnol.* **2012**, *4*, 219–233. [CrossRef] [PubMed]

45. Karpov, T.E.; Peltek, O.O.; Muslimov, A.R.; Tarakanchikova, Y.V.; Grunina, T.M.; Poponova, M.S.; Karyagina, A.S.; Chernozem, R.V.; Pariy, I.O.; Mukhortova, Y.R.; et al. Development of Optimized Strategies for Growth Factor Incorporation onto Electrospun Fibrous Scaffolds To Promote Prolonged Release. *ACS Appl. Mater. Interfaces* **2020**, *12*, 5578–5592. [CrossRef]

46. He, S.; Lin, K.F.; Fan, J.J.; Hu, G.; Dong, X.; Zhao, Y.N.; Song, Y.; Guo, Z.S.; Bi, L.; Liu, J. Synergistic Effect of Mesoporous Silica and Hydroxyapatite in Loaded Poly(DL-lactic-co-glycolic acid) Microspheres on the Regeneration of Bone Defects. *Biomed. Res. Int.* **2016**, *2016*, 9824827. [CrossRef] [PubMed]

47. Shuai, C.; Xu, Y.; Feng, P.; Xu, L.; Peng, S.; Deng, Y. Co-enhance bioactive of polymer scaffold with mesoporous silica and nano-hydroxyapatite. *J. Biomater. Sci. Polym. Ed.* **2019**, *30*, 1–17. [CrossRef] [PubMed]

48. Andrade, G.F.; Gomide, V.S.; Da Silva Junior, A.C.; Goes, A.M.; De Sousa, E.M. An in situ synthesis of mesoporous SBA-16/hydroxyapatite for ciprofloxacin release: In vitro stability and cytocompatibility studies. *J. Mater. Sci. Mater. Med.* **2014**, *25*, 2527–2540. [CrossRef]

49. Tamburaci, S.; Kimna, C.; Tihminlioglu, F. Bioactive diatomite and POSS silica cage reinforced chitosan/Na-carboxymethyl cellulose polyelectrolyte scaffolds for hard tissue regeneration. *Mater. Sci. Eng. C* **2019**. [CrossRef]

50. Gaihre, B.; Lecka-Czernik, B.; Jayasuriya, A.C. Injectable nanosilica-chitosan microparticles for bone regeneration applications. *J. Biomater. Appl.* **2017**. [CrossRef]

51. Khatami, N.; Khoshfetrat, A.B.; Khaksar, M.; Zamani, A.R.N.; Rahbarghazi, R. Collagen-alginate-nano-silica microspheres improved the osteogenic potential of human osteoblast-like MG-63 cells. *J. Cell. Biochem.* **2019**. [CrossRef]

52. Sun, J.L.; Jiao, K.; Song, Q.; Ma, C.F.; Ma, C.; Tay, F.R.; Niu, L.N.; Chen, J.H. Intrafibrillar silicified collagen scaffold promotes in-situ bone regeneration by activating the monocyte p38 signaling pathway. *Acta Biomater.* **2017**, *354–365*. [CrossRef]

53. Dai, C.; Guo, H.; Lu, J.; Shi, J.; Wei, J.; Liu, C. Osteogenic evaluation of calcium/magnesium-doped mesoporous silica scaffold with incorporation of rhBMP-2 by synchrotron radiation-based muCT. *Biomaterials* **2011**, *32*, 8506–8517. [CrossRef]

54. Peng, X.Y.; Hu, M.; Liao, F.; Yang, F.; Ke, Q.F.; Guo, Y.P.; Zhu, Z.H. La-Doped mesoporous calcium silicate/chitosan scaffolds for bone tissue engineering. *Biomater. Sci.* **2019**, *7*, 1565–1573. [CrossRef] [PubMed]

55. Shi, M.; Xia, L.; Chen, Z.; Lv, F.; Zhu, H.; Wei, F.; Han, S.; Chang, J.; Xiao, Y.; Wu, C. Europium-doped mesoporous silica nanosphere as an immune-modulating osteogenesis/angiogenesis agent. *Biomaterials* **2017**, *176–187*. [CrossRef] [PubMed]

56. Liao, F.; Peng, X.Y.; Yang, F.; Ke, Q.F.; Zhu, Z.H.; Guo, Y.P. Gadolinium-doped mesoporous calcium silicate/chitosan scaffolds enhanced bone regeneration ability. *Mater. Sci. Eng. C* **2019**, *104*, 109999. [CrossRef] [PubMed]
57. Wu, S.; Liu, X.; Yeung, K.W.K.; Liu, C.; Yang, X. Biomimetic porous scaffolds for bone tissue engineering. *Mater. Sci. Eng. R Rep.* **2014**, *80*, 1–36. [CrossRef]
58. Karageorgiou, V.; Kaplan, D. Porosity of 3D biomaterial scaffolds and osteogenesis. *Biomaterials* **2005**, *26*, 5474–5491. [CrossRef]
59. Dohan Ehrenfest, D.M.; Coelho, P.G.; Kang, B.S.; Sul, Y.T.; Albrektsson, T. Classification of osseointegrated implant surfaces: Materials, chemistry and topography. *Trends Biotechnol.* **2010**, *28*, 198–206. [CrossRef]
60. Ding, Y.; Yuan, Z.; Liu, P.; Cai, K.; Liu, R. Fabrication of strontium-incorporated protein supramolecular nanofilm on titanium substrates for promoting osteogenesis. *Mater. Sci. Eng. C* **2020**, *111*, 110851. [CrossRef]
61. Li, Q.; Wang, Z. Involvement of FAK/P38 Signaling Pathways in Mediating the Enhanced Osteogenesis Induced by Nano-Graphene Oxide Modification on Titanium Implant Surface. *Int. J. Nanomed.* **2020**, *15*, 4659–4676. [CrossRef]
62. Prabakaran, S.; Rajan, M.; Lv, C.; Meng, G. Lanthanides-Substituted Hydroxyapatite/Aloe vera Composite Coated Titanium Plate for Bone Tissue Regeneration. *Int. J. Nanomed.* **2020**, *15*, 8261–8279. [CrossRef]
63. Chen, L.; Mou, S.; Li, F.; Zeng, Y.; Sun, Y.; Horch, R.E.; Wei, W.; Wang, Z.; Sun, J. Self-Assembled Human Adipose-Derived Stem Cell-Derived Extracellular Vesicle-Functionalized Biotin-Doped Polypyrrole Titanium with Long-Term Stability and Potential Osteoinductive Ability. *ACS Appl. Mater. Interfaces* **2019**, *11*, 46183–46196. [CrossRef]
64. Zhang, B.; Li, J.; He, L.; Huang, H.; Weng, J. Bio-surface coated titanium scaffolds with cancellous bone-like biomimetic structure for enhanced bone tissue regeneration. *Acta Biomater.* **2020**, *114*, 431–448. [CrossRef] [PubMed]
65. Huang, Q.; Ouyang, Z.; Tan, Y.; Wu, H.; Liu, Y. Activating macrophages for enhanced osteogenic and bactericidal performance by Cu ion release from micro/nano-topographical coating on a titanium substrate. *Acta Biomater.* **2019**, *100*, 415–426. [CrossRef]
66. Zhao, S.; Xu, Y.; Xu, W.; Weng, Z.; Cao, F.; Wan, X.; Cui, T.; Yu, Y.; Liao, L.; Wang, X. Tremella-Like ZnO@Col-I-Decorated Titanium Surfaces with Dual-Light-Defined Broad-Spectrum Antibacterial and Triple Osteogenic Properties. *ACS Appl. Mater. Interfaces* **2020**, *12*, 30044–30051. [CrossRef]
67. Fu, X.; Liu, P.; Zhao, D.; Yuan, B.; Xiao, Z.; Zhou, Y.; Yang, X.; Zhu, X.; Tu, C.; Zhang, X. Effects of Nanotopography Regulation and Silicon Doping on Angiogenic and Osteogenic Activities of Hydroxyapatite Coating on Titanium Implant. *Int. J. Nanomed.* **2020**, *15*, 4171–4189. [CrossRef]
68. Long, M.; Rack, H.J. Titanium alloys in total joint replacement—A materials science perspective. *Biomaterials* **1998**, *19*, 1621–1639. [CrossRef]
69. Van Bael, S.; Chai, Y.C.; Truscello, S.; Moesen, M.; Kerckhofs, G.; Van Oosterwyck, H.; Kruth, J.P.; Schrooten, J. The effect of pore geometry on the in vitro biological behavior of human periosteum-derived cells seeded on selective laser-melted Ti6Al4V bone scaffolds. *Acta Biomater.* **2012**, *8*, 2824–2834. [CrossRef] [PubMed]
70. Vrancken, B.; Thijs, L.; Kruth, J.P.; Van Humbeeck, J. Microstructure and mechanical properties of a novel β titanium metallic composite by selective laser melting. *Acta Mater.* **2014**, *68*, 150–158. [CrossRef]
71. Wang, J.C.; Liu, Y.J.; Qin, P.; Liang, S.X.; Sercombe, T.B.; Zhang, L.C. Selective laser melting of Ti–35Nb composite from elemental powder mixture: Microstructure, mechanical behavior and corrosion behavior. *Mater. Sci. Eng.* **2019**, *760*, 214–224. [CrossRef]
72. Leong, S.S.; Edith, W.F.; Yee, Y.W. Selective laser melting of titanium alloy with 50 wt% tantalum: Effect of laser process parameters on part quality. *Int. J. Refract. Met. Hard Mater.* **2018**, *77*, 120–127.
73. Cho, H.R.; Roh, T.S.; Shim, K.W.; Kim, Y.O.; Lew, D.H.; Yun, I.S. Skull Reconstruction with Custom Made Three-Dimensional Titanium Implant. *Arch. Craniofac.Surg.* **2015**, *16*. [CrossRef] [PubMed]
74. Park, E.K.; Lim, J.Y.; Yun, I.S.; Kim, J.S.; Woo, S.H.; Kim, D.S.; Shim, K.W. Cranioplasty Enhanced by Three-Dimensional Printing: Custom-Made Three-Dimensional-Printed Titanium Implants for Skull Defects. *J. Craniofac. Surg.* **2016**, *27*, 943–949. [CrossRef]
75. Lee, Y.W.; You, H.J.; Jung, J.A.; Kim, D.W. Mandibular reconstruction using customized three-dimensional titanium implant. *Arch. Craniofac. Surg.* **2018**, *19*. [CrossRef] [PubMed]
76. Hamid, K.S.; Parekh, S.G.; Adams, S.B. Salvage of Severe Foot and Ankle Trauma With a 3D Printed Scaffold. *Foot Ankle Int.* **2016**, *37*, 433–439. [CrossRef]
77. Williams, L.R.; Fan, K.F.; Bentley, R.P. Custom-made titanium cranioplasty: Early and late complications of 151 cranioplasties and review of the literature. *Int. J. Oral Maxillofac. Surg.* **2015**, *44*, 599–608. [CrossRef] [PubMed]
78. Pei, X.; Wu, L.; Zhou, C.; Fan, H.; Gou, M.; Li, Z.; Zhang, B.; Lei, H.; Sun, H.; Liang, J.; et al. 3D printed titanium scaffolds with homogeneous diamond-like structures mimicking that of the osteocyte microenvironment and its bone regeneration study. *Biofabrication* **2020**. [CrossRef]
79. Maietta, S.; Gloria, A.; Improta, G.; Richetta, M.; Martorelli, M. A Further Analysis on Ti6Al4V Lattice Structures Manufactured by Selective Laser Melting. *J. Healthc. Eng.* **2019**, *2019*, 1–9. [CrossRef] [PubMed]
80. Zhu, M.; Ye, H.; Fang, J.; Zhong, C.; Yao, J.; Park, J.; Lu, X.; Ren, F. Engineering High-Resolution Micropatterns Directly onto Titanium with Optimized Contact Guidance to Promote Osteogenic Differentiation and Bone Regeneration. *ACS Appl. Mater. Interfaces* **2019**, *11*, 43888–43901. [CrossRef]
81. Saruta, J.; Sato, N.; Ishijima, M.; Okubo, T.; Hirota, M.; Ogawa, T. Disproportionate Effect of Sub-Micron Topography on Osteoconductive Capability of Titanium. *Int. J. Mol. Sci.* **2019**, *20*, 4027. [CrossRef] [PubMed]

82. Gorgin Karaji, Z.; Jahanmard, F.; Mirzaei, A.H.; Van der Wal, B.; Amin Yavari, S. A multifunctional silk coating on additively manufactured porous titanium to prevent implant-associated infection and stimulate bone regeneration. *Biomed. Mater.* **2020**, *15*, 065016. [CrossRef]

83. Sun, P.; Wang, Y.; Xu, D.; Gong, K. The Calcium Phosphate Modified Titanium Implant Combined With Platelet-Rich Plasma Treatment Promotes Implant Stabilization in an Osteoporotic Model. *J. Craniofac. Surg.* **2020**. [CrossRef]

84. Teng, F.Y.; Tai, I.C.; Ho, M.L.; Wang, J.W.; Weng, L.W.; Wang, Y.J.; Wang, M.W.; Tseng, C.C. Controlled release of BMP-2 from titanium with electrodeposition modification enhancing critical size bone formation. *Mater. Sci. Eng. C* **2019**, *105*, 109879. [CrossRef]

85. Hung, C.C.; Chaya, A.; Liu, K.; Verdelis, K.; Sfeir, C. The role of magnesium ions in bone regeneration involves the canonical Wnt signaling pathway. *Acta Biomater.* **2019**, *98*, 246–255. [CrossRef]

86. Hamushan, M.; Cai, W.; Zhang, Y.; Lou, T.; Zhang, S.; Zhang, X.; Cheng, P.; Zhao, C.; Han, P. High-purity magnesium pin enhances bone consolidation in distraction osteogenesis model through activation of the VHL/HIF-1alpha/VEGF signaling. *J. Biomater. Appl.* **2020**, *35*, 224–236. [CrossRef] [PubMed]

87. Zhao, D.; Witte, F.; Lu, F.; Wang, J.; Li, J.; Qin, L. Current status on clinical applications of magnesium-based orthopaedic implants: A review from clinical translational perspective. *Biomaterials* **2017**, *112*, 287–302. [CrossRef] [PubMed]

88. Tie, D.; Guan, R.; Liu, H.; Cipriano, A.; Liu, Y.; Wang, Q.; Huang, Y.; Hort, N. An in vivo study on the metabolism and osteogenic activity of bioabsorbable Mg-1Sr alloy. *Acta Biomater.* **2016**, *29*, 455–467. [CrossRef]

89. Niu, J.; Xiong, M.; Guan, X.; Zhang, J.; Huang, H.; Pei, J.; Yuan, G. The in vivo degradation and bone-implant interface of Mg-Nd-Zn-Zr alloy screws: 18 months post-operation results. *Corros. Sci.* **2016**, *113*, 183–187. [CrossRef]

90. Santos-Coquillat, A.; Esteban-Lucia, M.; Martinez-Campos, E.; Mohedano, M.; Arrabal, R.; Blawert, C.; Zheludkevich, M.L.; Matykina, E. PEO coatings design for Mg-Ca alloy for cardiovascular stent and bone regeneration applications. *Mater. Sci. Eng. C* **2019**, *105*, 110026. [CrossRef] [PubMed]

91. Zhang, D.; Ni, N.; Su, Y.; Miao, H.; Tang, Z.; Ji, Y.; Wang, Y.; Gao, H.; Ju, Y.; Sun, N.; et al. Targeting Local Osteogenic and Ancillary Cells by Mechanobiologically Optimized Magnesium Scaffolds for Orbital Bone Reconstruction in Canines. *ACS Appl. Mater. Interfaces* **2020**, *12*, 27889–27904. [CrossRef]

92. Perumal, G.; Ramasamy, B.; Nandkumar, A.M.; Dhanasekaran, S.; Ramasamy, S.; Doble, M. Bilayer nanostructure coated AZ31 magnesium alloy implants: In vivo reconstruction of critical-sized rabbit femoral segmental bone defect. *Nanomedicine* **2020**, *29*, 102232. [CrossRef]

93. Lin, Z.; Zhao, Y.; Chu, P.K.; Wang, L.; Pan, H.; Zheng, Y.; Wu, S.; Liu, X.; Cheung, K.M.C.; Wong, T.; et al. A functionalized TiO2/Mg2TiO4 nano-layer on biodegradable magnesium implant enables superior bone-implant integration and bacterial disinfection. *Biomaterials* **2019**, *219*, 119372. [CrossRef]

94. Bessa-Goncalves, M.; Silva, A.M.; Bras, J.P.; Helmholz, H.; Luthringer-Feyerabend, B.J.C.; Willumeit-Romer, R.; Barbosa, M.A.; Santos, S.G. Fibrinogen and magnesium combination biomaterials modulate macrophage phenotype, NF-kB signaling and crosstalk with mesenchymal stem/stromal cells. *Acta Biomater.* **2020**, *114*, 471–484. [CrossRef]

95. Li, G.; Zhang, L.; Wang, L.; Yuan, G.; Dai, K.; Pei, J.; Hao, Y. Dual modulation of bone formation and resorption with zoledronic acid-loaded biodegradable magnesium alloy implants improves osteoporotic fracture healing: An in vitro and in vivo study. *Acta Biomater.* **2018**, *65*, 486–500. [CrossRef]

96. Salehi, M.; Maleksaeedi, S.; Sapari, M.A.B.; Nai, M.L.S.; Meenashisundaram, G.K.; Gupta, M. Additive manufacturing of magnesium–zinc–zirconium (ZK) alloys via capillary-mediated binderless three-dimensional printing—ScienceDirect. *Mater. Des.* **2019**, *169*, 107683. [CrossRef]

97. Li, Y.; Zhou, J.; Pavanram, P.; Leeflang, M.A.; Fockaert, L.I.; Pouran, B.; Tumer, N.; Schroder, K.U.; Mol, J.M.C.; Weinans, H.; et al. Additively manufactured biodegradable porous magnesium. *Acta Biomater.* **2018**, *67*, 378–392. [CrossRef]

98. Zhu, D.; Su, Y.; Young, M.L.; Ma, J.; Zheng, Y.; Tang, L. Biological Responses and Mechanisms of Human Bone Marrow Mesenchymal Stem Cells to Zn and Mg Biomaterials. *ACS Appl. Mater. Interfaces* **2017**, *9*, 27453–27461. [CrossRef]

99. Bowen, P.K.; Drelich, J.; Goldman, J. Zinc exhibits ideal physiological corrosion behavior for bioabsorbable stents. *Adv. Mater.* **2013**, *25*, 2577–2582. [CrossRef]

100. Zhang, Y.; Yan, Y.; Xu, X.; Lu, Y.; Chen, L.; Li, D.; Dai, Y.; Kang, Y.; Yu, K. Investigation on the microstructure, mechanical properties, in vitro degradation behavior and biocompatibility of newly developed Zn-0.8%Li-(Mg, Ag) alloys for guided bone regeneration. *Mater. Sci. Eng. C* **2019**, *99*, 1021–1034. [CrossRef]

101. Montani, M.; Demir, A.G.; Mostaed, E.; Vedani, M.; Previtali, B. Processability of pure Zn and pure Fe by SLM for biodegradable metallic implant manufacturing. *Rapid Prototyp. J.* **2017**. [CrossRef]

102. Li, Y.; Pavanram, P.; Zhou, J.; Lietaert, K.; Bobbert, F.S.L.; Kubo, Y.; Leeflang, M.A.; Jahr, H.; Zadpoor, A.A. Additively manufactured functionally graded biodegradable porous zinc. *Biomater. Sci.* **2020**, *8*, 2404–2419. [CrossRef] [PubMed]

103. Lietaert, K.; Zadpoor, A.A.; Sonnaert, M.; Schrooten, J.; Weber, L.; Mortensen, A.; Vleugels, J. Mechanical properties and cytocompatibility of dense and porous Zn produced by laser powder bed fusion for biodegradable implant applications. *Acta Biomater.* **2020**, *110*, 289–302. [CrossRef]

104. Jakhmola, A.; Vecchione, R.; Gentile, F.; Profeta, M.; Manikas, A.C.; Battista, E.; Celentano, M.; Onesto, V.; Netti, P.A. Experimental and theoretical study of biodirected green synthesis of gold nanoflowers. *Mater. Today Chem.* **2019**, *14*. [CrossRef]

105. Jakhmola, A.; Vecchione, R.; Onesto, V.; Gentile, F.; Profeta, M.; Battista, E.; Manikas, A.C.; Netti, P.A. A theoretical and experimental study on L-tyrosine and citrate mediated sustainable production of near infrared absorbing twisted gold nanorods. *Mat. Sci. Eng. C* **2021**, *118*. [CrossRef] [PubMed]
106. Celentano, M.; Jakhmola, A.; Profeta, M.; Battista, E.; Guarnieri, D.; Gentile, F.; Netti, P.A.; Vecchione, R. Diffusion limited green synthesis of ultra-small gold nanoparticles at room temperature. *Colloid Surf. A* **2018**, *558*, 548–557. [CrossRef]
107. Russo, T.; Gloria, A.; De Santis, R.; D'Amora, U.; Balato, G.; Vollaro, A.; Oliviero, O.; Improta, G.; Triassi, M.; Ambrosio, L. Preliminary focus on the mechanical and antibacterial activity of a PMMA-based bone cement loaded with gold nanoparticles. *Bioact. Mater.* **2017**, *2*, 156–161. [CrossRef]
108. Liao, J.; Shi, K.; Jia, Y.; Wu, Y.; Qian, Z. Gold nanorods and nanohydroxyapatite hybrid hydrogel for preventing bone tumor recurrence via postoperative photothermal therapy and bone regeneration promotion. *Bioact. Mater.* **2021**, *6*, 2221–2230. [CrossRef]
109. Shi, M.; Zhang, P.; Zhao, Q.; Shen, K.; Qiu, Y.; Xiao, Y.; Yuan, Q.; Zhang, Y. Dual Functional Monocytes Modulate Bactericidal and Anti-Inflammation Process for Severe Osteomyelitis Treatment. *Small* **2020**, *16*, e2002301. [CrossRef]
110. Sanchez-Casanova, S.; Martin-Saavedra, F.M.; Escudero-Duch, C.; Uceda, M.I.F.; Prieto, M.; Arruebo, M.; Acebo, P.; Fabiilli, M.L.; Franceschi, R.T.; Vilaboa, N. Local delivery of bone morphogenetic protein-2 from near infrared-responsive hydrogels for bone tissue regeneration. *Biomaterials* **2020**, *241*. [CrossRef]

 *nanomaterials*

Article

Nano-Hydroxyapatite vs. Xenografts: Synthesis, Characterization, and In Vitro Behavior

Cristina Rodica Dumitrescu [1], Ionela Andreea Neacsu [1,2,*], Vasile Adrian Surdu [1,2], Adrian Ionut Nicoara [1,2], Florin Iordache [3], Roxana Trusca [2], Lucian Toma Ciocan [4], Anton Ficai [1,2,5] and Ecaterina Andronescu [1,2,5]

[1] Department of Science and Engineering of Oxide Materials and Nanomaterials, Faculty of Applied Chemistry and Materials Science, University Politehnica of Bucharest, 060042 Bucharest, Romania; cristinadumitrescu0@gmail.com (C.R.D.); adrian.surdu@upb.ro (V.A.S.); adrian.nicoara@upb.ro (A.I.N.); anton.ficai@upb.ro (A.F.); ecaterina.andronescu@upb.ro (E.A.)

[2] National Research Center for Micro and Nanomaterials, Faculty of Applied Chemistry and Materials Science, University Politehnica of Bucharest, 060042 Bucharest, Romania; truscaroxana@yahoo.com

[3] Department of Biochemistry, Faculty of Veterinary Medicine, University of Agronomic Science and Veterinary Medicine, 011464 Bucharest, Romania; floriniordache84@yahoo.com

[4] Prosthetics Technology and Dental Materials Department, Carol Davila University of Medicine and Pharmacy, 020022 Bucharest, Romania; tciocan@yahoo.com

[5] National Research Center for Food Safety, University Politehnica of Bucharest, 060042 Bucharest, Romania

* Correspondence: ionela.neacsu@upb.ro

Citation: Dumitrescu, C.R.; Neacsu, I.A.; Surdu, V.A.; Nicoara, A.I.; Iordache, F.; Trusca, R.; Ciocan, L.T.; Ficai, A.; Andronescu, E. Nano-Hydroxyapatite vs. Xenografts: Synthesis, Characterization, and In Vitro Behavior. *Nanomaterials* **2021**, *11*, 2289. https://doi.org/10.3390/nano11092289

Academic Editors: Saso Ivanovski, Karan Gulati and Abdalla Ali

Received: 29 July 2021
Accepted: 29 August 2021
Published: 2 September 2021

Publisher's Note: MDPI stays neutral with regard to jurisdictional claims in published maps and institutional affiliations.

Abstract: This research focused on the synthesis of apatite, starting from a natural biogenic calcium source (egg-shells) and its chemical and morpho-structural characterization in comparison with two commercial xenografts used as a bone substitute in dentistry. The synthesis route for the hydroxyapatite powder was the microwave-assisted hydrothermal technique, starting from annealed egg-shells as the precursor for lime and di-base ammonium phosphate as the phosphate precursor. The powders were characterized by Fourier-transform infrared spectroscopy (FTIR), X-ray diffraction (XRD), scanning electron microscopy (SEM), energy-dispersive X-ray analysis (EDAX), transmission electron microscopy (TEM), X-ray fluorescence spectroscopy (XRF), and cytotoxicity assay in contact with amniotic fluid stem cell (AFSC) cultures. Compositional and structural similarities or differences between the powder synthesized from egg-shells (HA1) and the two commercial xenograft powders— Bio-Oss®, totally deproteinized cortical bovine bone, and Gen-Os®, partially deproteinized porcine bone—were revealed. The HA1 specimen presented a single mineral phase as polycrystalline apatite with a high crystallinity (X_c 0.92), a crystallite size of 43.73 nm, preferential growth under the c axes (002) direction, where it mineralizes in bone, a nano-rod particle morphology, and average lengths up to 77.29 nm and diameters up to 21.74 nm. The surface of the HA1 nanoparticles and internal mesopores (mean size of 3.3 ± 1.6 nm), acquired from high-pressure hydrothermal maturation, along with the precursor's nature, could be responsible for the improved biocompatibility, biomolecule adhesion, and osteoconductive abilities in bone substitute applications. The cytotoxicity assay showed a better AFSC cell viability for HA1 powder than the commercial xenografts did, similar oxidative stress to the control sample, and improved results compared with Gen-Os. The presented preliminary biocompatibility results are promising for bone tissue regeneration applications of HA1, and the study will continue with further tests on osteoblast differentiation and mineralization.

Keywords: biomaterial; bone substitute; apatite; microwave-assisted hydrothermal synthesis

1. Introduction

After blood transfusions, bone grafts are placed on the second position as the most frequent tissue transplantation used worldwide [1]. Multiple factors determine the occurrence of alveolar bone defects, but the most common are osseous deficiency, the resection of tumors, alveolar bone loss due to periodontal disease, and subsequent tooth loss [2]. Bone defects require rehabilitation, mainly to avoid severe alveolar bone resorption that compromises bone quantity, morphology, and quality, which prevents the failure of dental implant placement, the maintaining of the normal anatomic outline, the elimination of empty space, aesthetic restoration, as well as drugs encapsulation and delivery [3–5]. Until now, only the autografts and, partially, allografts could fulfil the desired properties such as biocompatibility, osteoconductivity, and osteoinductivity for a better bone substitution ability, albeit entailing important risks [6–10]. Xenografts (mammalian bone source) have been used for more than thirty years and are still being used with good clinical results, osteoconductive features, very good biocompatibility, high availability (size and quantity), and low cost, but also involving many disadvantages [3,11–19].

Therefore, natural sources for biological apatite synthesis have been extended, year by year, from mammalian bone sources next to fish bones and scales [20–22], egg-shells [23,24], and exoskeletons of marine organisms (snails, starfish, coral, and seashell) [25–28] or botanic sources (Calendula flower, Papaya leaf, and orange peel), where all of them need chemical and thermal preparation before use as a mammalian xenograft. However, the interest in this topic is still present, a fact proved by several recent publications [23–25]. Mammalian bone, after thermal treatments, provides so-called biological apatite [3,11–19]; egg-shells (hen, crocodile etc.) and marine organism skeletons, including corals and starfish, are special sources of biogenic calcium carbonate [23–28]; phytogenic calcium carbonates could be used after the extraction process in the original form, preserving the porous structure, calcium-rich source, and biomolecules (Ovalbumin, Carotene, Papain, and vitamins) [10,11].

The great advantages of alloplastic bone substitute consist of a wide range of compositions, sizes, shapes, textures, synthesis methods, biocompatibility and bioresorbability, high quantities of availability, and cost-effectiveness [29–32]. Despite these advantages, they still need improvements in order to enhance their integration into the physiological environment, their bioactivity, or biotolerance [33–35].

Nowadays, one of the most promising approaches is hydroxyapatite synthesis from hen egg-shell sources due to the mimetic composition and structure of the carbonated apatite obtained compared with human bone [36–38]. Such carbonated apatite is prepared using several routes, including dry methods (solid state [39] and mechanochemical synthesis [40]), wet methods (precipitation [41–43], hydrolysis, sol–gel [44], emulsion [45], sonochemical [46], hydrothermal [47,48], and solvothermal [49,50]), or high-temperature methods (pyrolysis, combustion, and microwave heating [50]). Each of the abovementioned methods induces different morphologies of micro- and nano-size HA particles: rods [51,52], sheets [53], spheres [54], wires [55], fibers [56], flower, worm, hexagonal prism, platelet, lath, strip, dandelion, chrysanthemum, rosette, or spheres [20]. Among those, the hydrothermal method generates conditions for high crystallinity and stoichiometric hydroxyapatite synthesis (Ca/P ratio around 1.67) from shells of calcium carbonate precursor, usually with a rod-like morphology and hexagonal unit cell symmetry [57,58]. Moreover, combining hydrothermal synthesis (HT) with microwave thermal treatment (MW) was observed to lead to a higher crystal size of HA from egg-shells after only 1–36 min for MW-HT, compared to several hours when using HT alone; a higher pH (between 9–11) and reaction time generate a high content of carbonate groups in the resulting HA lattice [53,59–61], but the high pressure in the synthesis system ensures a high internal and external porosity, as has been reported only in few papers [60]. Besides, a wide range of morphologies were obtained only using the two combined techniques for nano-hydroxyapatite synthesis. Furthermore, the association between the two techniques, HT and MW, was reported to improve the control of particle size, porosity, and morphology by a better monitoring of process parameters (time, temperature, and energy), with low energy consumption, low temperatures (less than 250 °C), and short time process cycles [51–61].

In this paper, we compared three categories of biomaterials obtained from natural sources, two represented by Bio-Oss®—bovine bone and Gen-OS®—porcine bone, bone grafts already used in dentistry with good clinical results [18,62–64], and the last one being biomimetic synthetic hydroxyapatite from egg-shells (HA1), synthesized by the microwave-assisted hydrothermal technique (HT-MW), after only two hours of treatment at 200 °C. The usage of this unconventional, hybrid synthesis method with microwave heating has already been reported in the literature with good results. The fact that a natural calcium source has been proposed as a starting material aims to bring added value to this work and proposes an alternative to the actual, expensive, commercial materials. Moreover, the paper is intended as an extensive comparative study between the synthetic material and the two bone grafts already used in dentistry, highlighting their chemical, structural, morphological, and biological resemblance.

2. Materials and Methods

2.1. Materials

Gen-Oss® powder was purchased from Tecnoss Dental (Pianezza, Italy). It is a mixture of grinded cortical (20% wt.) and cancellous (80% wt.) porcine bone, obtained by low-temperature treatment (maximum 130 °C) in order to partially burn out the organic compound of bone, the version with grain sizes of 250–1000 μm.

Bio-Oss® powder was purchased from Geistlich Pharma AG (Wolhusen, Switzerland). It is a chemically and thermally treated cancellous bovine bone. Hence, grinded bovine bone was first deproteinized by reaction with the strong alkali medium, and then calcined at 300 °C, the version available with 0.25–1 mm grain sizes.

Hydroxyapatite (HA) was synthesized from a natural calcium carbonate source—hen egg-shells.

2.1.1. Calcium Oxide Precursor Preparation

Here, 100 g of hen eggshells were harvested from a local poultry and boiled for 4 h in water with 10 mL of H_2O_2, for complete removal of the organic part, and then washed with distilled water, dried in an oven at 60 °C, and ground for 15 min. The ground material was annealed in an electric oven, with a temperature rise rate of 10 °C/min, up to a temperature of 800 °C, bearing 3 h, and then slowly cooled to ambient temperature for 24 h.

2.1.2. Hydroxyapatite Preparation

Here, 45 g of annealed egg-shell powder was dispersed in 200 mL of distilled water and further used as a Ca source. Then, 100 mL of aqueous solution of 38.5% of di-base ammonium phosphate $[(NH_4)_2HPO_4]$ was prepared and used as a phosphorus precursor. The two reagents were mixed under continuous magnetic stirring, by dripping di-base ammonium phosphate over a Ca source at an average speed of 2 mL/min, periodically adjusting the pH > 11. This value has been reported in the literature to favor the formation of rod-like hydroxyapatite [30]. The precipitate obtained was maturated in a microwave-assisted hydrothermal Teflon autoclave.

2.1.3. Microwave-Assisted Hydrothermal Maturation of Hydroxyapatite Precipitate

The precipitate was introduced into a 50 cm^3 Teflon vessel with an occupancy rate of 50% [59], being subjected to a hydrothermal-microwave heating treatment as follows: the temperature increased from room temperature to 200 °C at a rate of 35 °C/min, where it was maintained for 30 min, and then slowly decreased back to room temperature. Throughout the heating cycle, the MW energy supplied to the system varied in the first 5 min of treatment between 1.2 and 1.6 kW and, during the 30 min at maximum temperature, the range was <1.0 kW. The pressure in the system increased during the hydrothermal maturation in the first 40 min, from the initial input value of 16 bar at approximately 20 bar, remaining around this value also during the cooling period. The maturated precipitate was filtered and washed with distilled water until pH = 7, and then dried at 60 °C for 48 h, resulting in HA1 powder.

2.2. Characterisation Methods

X-ray diffraction (XRD) was performed using a PANalytical Empyrean Spectrometer (Malvern PANalytical, Bruno, The Netherlands), operating in a Bragg-Brentano configuration with Cu-Kα (λ = 1.5406 Å). The spectra were recorded at $100 < 2\theta < 80°$ with a scan speed of 0.5°/min and a step size of 0.02°. Using the following empirical equation, Equation (1), the crystallinity degree for every powder diffraction pattern can be appreciated [46]:

$$Xc = 1 - \frac{V112/300}{I\,300} \tag{1}$$

where I_{300} is the intensity of the reflection crystal plane (300) and $V_{112/300}$ is the intensity of the difference between (112) and (300) reflections (which completely disappears in noncrystalline samples of hydroxyapatite) [65–67].

In order to calculate the average hydroxyapatite crystallite size of Bio-Oss, Gen-OS, and HA1 powders, the Rietveld method was applied, based on all X-ray diffraction peak profiles in the pattern, using HighScorePlus 3.0.e software and the pseudo-Voigt function for the profile refinement procedure [65–67]. The Scherer formula (2) was used to estimate the crystallite size only along the growth plane direction using a convolution of the Cauchy–Lorentz probability distribution that is marked in the XRD diagram by an increasing peak associated with a higher crystallinity degree of the powders [61]:

$$r = \frac{K\lambda}{B\,\cos\theta} \tag{2}$$

where r = crystallite size [nm], K = 0.9 constant [61], λ = wavelength of monochromatic X-ray beam [nm] ($\lambda\,K\alpha Cu = 0.15418$ nm), B is the full-width of the peak at half-intensity of each crystalline plane reflection [rad], and θ is the exact diffraction angle [rad].

Fourier-transform infrared spectroscopy (FTIR) spectra were recorded in the wavenumber range of 4000–500 cm^{-1}, in increments of 1.928 cm^{-1}, using a Nicolet iS50R spectrometer (Thermo Fisher, Waltham, MA, USA), in attenuated total reflection mode (ATR). Each spectrum was collected at room temperature at a resolution of 4 cm^{-1}, and 32 samples were scanned between 4000 and 440 cm^{-1}. The obtained results were presented as the average of the 32 individual scanned samples for each Bio-Oss, GenOs, and HA1 powder and were compared with the theoretical available data [68–71].

A Quanta Inspect F scanning electron microscope (SEM) (Thermo Fisher, Eindhoven, the Netherlands), equipped with a field electron emission gun (FEG) and an EDS (energy-dispersive spectroscopy) detector, was used. The technical parameters were: acceleration voltage of 30 kV and point-to-point resolution of 1.2 nm.

A TECNAI F 30G2 SWIN transmission electron microscope (TEM) (Thermo Fisher, Eindhoven, the Netherlands) was used, with a 300 kV acceleration transmission with a Shottky electron emission, HRTEM point and line resolutions of 2 Å and 1.02 Å, respectively, 60x-1Mx magnification range, and a minimum diffraction angle of $\pm12°$, equipped with an EDS probe.

The metal contents of both HA1 powder and egg-shell raw material were determined by X-ray fluorescence spectra (XRF) using a Thermo Scientific ARL PERFORM'X Sequential spectrometer, which works under pressure in the He atmosphere, and the purchase was made according to the Thermo Scientific UniQuant soft, nonstandard method.

In vitro qualitative biocompatibility was performed on a GM0047 amniotic fluid stem cell line (AFSC), purchased from Coriell Institute (Kenton, NJ, USA) and cultivated at the Faculty of Veterinary Medicine, Department of Biochemistry (Bucharest, Romania). The cells were cultivated in Dulbecco's modified Eagle's medium (DMEM, Sigma-Aldrich, Missouri, MO, USA) supplemented with 10% fetal bovine serum and 1% antibiotics (penicillin and streptomycin) and changed twice a week. The AFSC cell culture was obtained at a final concentration of 5 µM and RED CMTPX was added as the cell trace fluorophore. The cells were treated with Bio-Oss, Gen-Os, and HA1 granular materials and incubated for 30 min, allowing the chromophore penetration into the cells. The viability and morphology of the

AFSCs were appreciated after 5 days. The cell phenotype was evaluated by flow cytometry using specific markers such as SSEA-1, SSEA-4, TRA1–60, TRA 1–81, CD90, CD73, CD56, CD49E, CD44, CD31, CD105, and CD45. There was no modification in the cell phenotype after 5 days in the presence of the biomaterials [72,73]. After this period, in order to observe the cells' fluorescence, the AFSC medium was washed with phosphate-buffered saline (PBS) (8.0 g/L of NaCl, 0.2 g/L of KCl, 1.42 g/L of Na_2HPO_4, and 0.24 g/L of KH_2PO_4, pH~7.4). The micrographs were made with a digital camera Olympus CKX 41, driven by CellSense Entry software (Olympus, Tokyo, Japan).

Two quantitative evaluations of the in vitro cellular bioactivity in contact with the three biomaterials were performed: Viability and Oxidative Stress Assessments (MTT and GSH-Glo Glutathione Assays). The GSH-Glo Assay is based on the conversion of a luciferin derivative (Luc-NT-luciferin dimer) coupled in the presence of glutathione (GSH), oxygen, enzymes (luciferase and glutation S transferase), and ATP. Adding the marker of luciferase Ultra-Glo Recombinant Luciferase is necessary to produce luminescence, which is proportional to the quantity of GSH produced in cells. GSH is the main thiol of animal cells involved in important metabolic mechanisms such as signaling biomolecules in redox reactions, the regulation of cell proliferation, and fibrogenesis, and its level is a measure of the antioxidative stress of cells [74]. AFSCs were seeded for 24 h, at a density of 3000 cells in 300 µL of DMEM supplemented with 10% fetal bovine serum and 1% antibiotics (penicillin and streptomycin/neomycin) in 96-well plates. After preparing the cell culture, the tested biomaterials (Bio-Oss, Gen-Os, and HA1) were put in contact and incubated for 72 h. The protocol involves the addition of 100 µL of 1X GSH-Glo Reagent and incubating it at 37 °C for 30 min, followed by adding 100 µL of Luciferin Detection Reagent for another 15 min, also incubated at 37 °C. Three independent wells of each sample were observed on a luminometer (Microplate Luminometer Centro LB 960, Berthold, Germany), after a good homogenization. The change in density, produced by solubilized formazan, was appreciated spectrophotometrically (TECAN Infinite M200, Männedorf, Switzerland) (Thermo Fischer Scientific, Waltham, MA, USA); the absorbance of solubilized formazan concentration is proportional to the metabolic activity of living cells in the culture. The human mesenchymal amniotic fluid stem cells (AFSCs) (Vybrant®MTT Cell Proliferation Assay Kit) were cultured in 96-well plates, with a seed density of 3000 cells/well, in the presence of analyzed Bio-Oss, Gen-Os, and HA1 powders, into DMEM medium (Sigma-Aldrich, Saint Luis, MI, USA) with the addition of 10% fetal bovine serum, 1% penicillin, and 1% streptomycin antibiotics (Sigma-Aldrich, Saint Luis, MI, USA), for 72 h. After the incubation period, 15 mL of MTT (12 mM) was added and kept for 4 h at 37 °C, and using a pipette, the solution of 1 mg of sodium dodecyl sulphate + 10 mL of HCl and 0.01 M was appended to the formazan crystals solubilization. The absorbance was measured after 1 h, in triplicate, using a spectrophotometer at 570 nm [75–78].

3. Results and Discussions

The synthesis strategy for the HT-MW maturation treatment was to control the process parameters in order to obtain a nanosize single-phase hydroxyapatite powder, with particles presenting mesopores. The usage of microwave radiation leads to a higher reaction speed, due to the polarization of water from the aqueous suspension [20]. Consequently, the treatment duration was reduced to only 30 min and the entire process took place with energy consumption savings. The maximum temperature of the process was settled at 200 °C in order to avoid the formation of secondary phosphate phases, as reported in the literature [19,20]. A homogeneous nanosize powder can be acquired by using the precipitation and hydrothermal synthesis methods [24,27], but taking care that the treatment time and temperature do not exceed the conditions for crystal growth velocity [20]. The rod-like morphology of the synthesized hydroxyapatite particles have been reported [52–54], even using the solvothermal-MW method, with different organic phosphate precursors [49–52], and the suspension pH seems to be the decisive parameter [57]. Therefore, the pH was carefully kept, during precipitate development, at a high basicity level over 11 by choosing a suitable regent. Adopting a high-pressure hydrothermal process (initial 16 bars) combined with gaseous reaction products, a high porosity of crystals was expected.

Figure 1 (left) represents the XRD pattern of the natural raw material (hen egg-shells) used for hydroxyapatite synthesis, before calcination, while Figure 2 comparatively presents the diffraction patterns for all three apatite materials. The XRD pattern in Figure 1 was matched by the calcite ($CaCO_3$) phase with a high degree of crystallinity evidenced by sharp diffraction peaks. The results are in good agreement with numerous references, which present egg-shells as a calcium carbonate source [24,25]. In addition, a calcium content of 96.38 wt.% was observed after XRF examination, made on egg-shells before calcination.

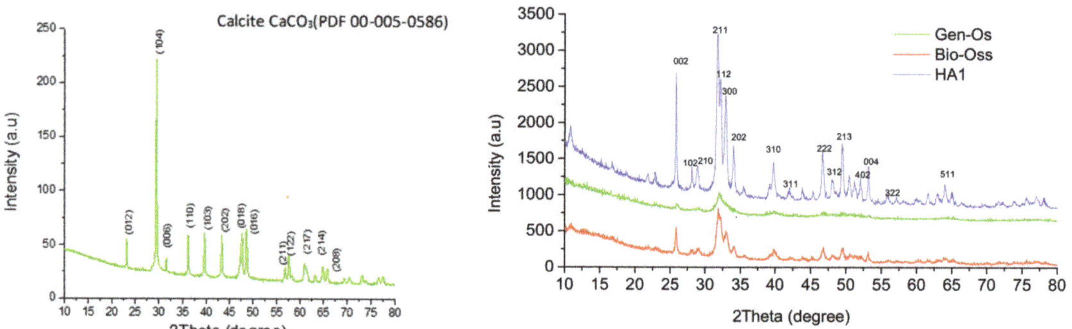

Figure 1. XRD plot for: (**left**) hen egg-shell powder before calcination; (**right**) HA1, Bio-Oss, and Gen-Os powders.

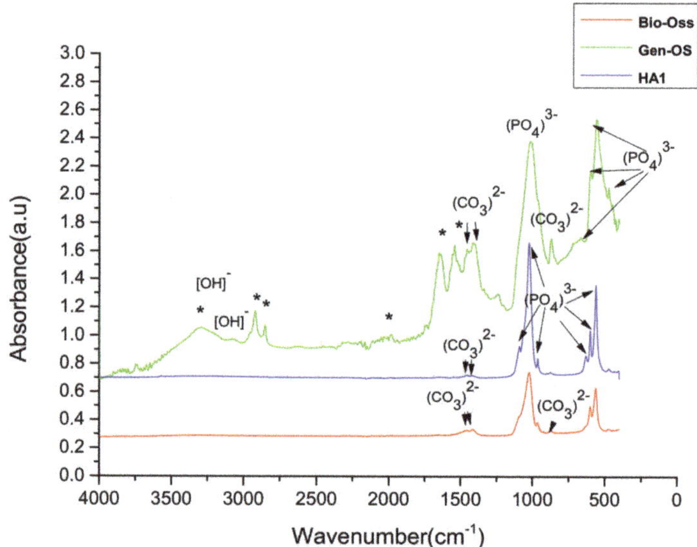

Figure 2. FTIR absorption spectra for xenograft Bio-Oss, Gen-Os, and HA1 powder: (*) absorption bands for protein functional groups.

According to the reference sheet PDF code 00-009-0432, the sample HA1 consists of 100% HA, and all characteristic crystalline planes were present with amplitude and diffraction angles (2θ) corresponding to a hexagonal symmetry hydroxyapatite pattern, as seen in Figure 1 (right, blue). No other secondary phases were present in the HA1 sample. For Bio-Oss® (Figure 1 right, red), the following main characteristic crystalline planes of HA were distinguished: (211) with a maximum amplitude (100%) at 2θ of 31.77°, followed by (112) and (300) of 60% at 2θ = 32.19° and 32.90°, respectively, (002) at 25.87° (40%), (310) at 2θ of 39.80° and (222) at 46.79°, (213) (18%) to 49.46°, and (004) to 53.14°. Gen-Os (Figure 1 right, green) proved to be the least crystallized sample of the three, as only the crystalline phase planes (002), (211), (112), and (310) of HA were identified. Moreover, Gen-Os registered a much smaller intensity and broad X-ray scattering profile at low diffraction angles, compared with the other two samples. As Gen-Os originated from porcine bone after thermal treatment at low temperature, the decrease in crystallinity can be attributed to the possible small numbers of remaining organic components, an hypothesis that was later confirmed by FTIR analysis (Figure 2). Compared to Gen-Os, Bio-Oss was better-crystallized, but the exhibited peaks corresponding to the crystalline planes of HA were smaller in intensity compared to the HA1 powder XRD pattern. From XRD patterns, applying Equation (1), the crystallinity degree was calculated, proving the highest crystallinity for HA1 (XC = 0.92), followed by Bio-Oss (XC = 0.56) and Gen-Os XC = 0.28 [46].

To determine the crystallinity degree and crystallite sizes of hydroxyapatite corresponding to the three samples of bone substitutes, the Rietveld method and Scherer equation were used [65–67]. To calculate the average size of crystallites, the Rietveld method is more precise compared with the Scherer equation, because it takes into consideration all peak separations or the total integrated intensity of groups of overlapping peaks from the diffraction plot, using a pseudo-Voigt function for a better matching profile of X-ray diffraction peaks. Using each of the three XRD patterns with the application of the Rietveld method, the crystallite average size of HA in the HA1 sample was the largest of the three samples (21.62 nm), almost double compared to Bio-Oss and almost triple compared to Gen-Os (Table 1), as already suggested by the small width of the HA1 XRD peaks presented in Figure 1 [65–67].

Table 1. Crystallite average size for HA1, Gen-Os, and Bio-Oss by Rietveld method.

Sample	Crystallite Average Size (nm)	Standard Deviation Value
Bio-Oss	12.65	1.45
Gen OS	7.52	0.89
HA1	21.62	1.89

Table 2 shows the calculated crystallite sizes for [002], [211], and [300] planes, having the highest diffraction peaks in each plot for the three samples HA1, Gen-Os, and Bio-Oss. Even though the (211) plane is associated with the highest-intensity peak for all samples, its corresponding crystallite sizes are not largest. It can be observed that the biggest crystallite growth occurs under the (002) plane for samples HA1 and Bio-Oss, of 23.44 nm and 43.73 nm, respectively, which was found in the literature to be typical for natural bone, where the *c* axes are the growth direction for collagen fibrils [68]. The Gen-Os sample shows the smallest crystallite sizes, including under the [002] direction, explained by the lowest amplitude of all diffraction peaks being registered for this sample (Figure 2 green), perhaps because small hydroxyapatite crystals are shielded by the wrapping of proteins. The results obtained through both calculation techniques are comparable and present the same order for increasing the mean crystallite size of Gen-Os < Bio-Oss < HA1.

Table 2. Crystallite size by Scherer equation for the highest crystalline planes.

Sample	FWHM	2 θ (°)	Crystalline Direction	Crystallite Size (nm)	Mean Crystallite Size (nm)
Bio-Oss	0.347	25.88	[002]	23.44	
	0.856	31.86	[211]	9.64	14.69
	0.753	32.93	[300]	10.99	
Gen-Os	3.874	25.88	[002]	2.10	
	1.351	31.99	[211]	6.13	5.45
	1.017	33.10	[300]	8.14	
HA1	0.186	25.78	[002]	43.73	
	0.752	31.87	[211]	10.98	23.83
	0.447	33.10	[300]	18.53	

The FTIR spectra in Figure 2 are attributed to HA1 (blue line), Bio-Oss (red line), and Gen-Os (green line) powders. Symmetric vibration bending or stretching of the absorption bands can be observed for the C-O bond at wavenumbers of 1454, 1420, and 874 cm^{-1}, which can be attributed to CO_3^{2-} (carbonate ions type B) that substitutes PO_4^{3-} in the HA lattice. The PO_4^{3-} groups absorption bands identified at 472 cm^{-1} are characteristic of the ν$_2$ inclination of the O-P-O bond, the high-amplitude bands from 572 cm^{-1}, 601 cm^{-1}, and 963 cm^{-1} refer to the symmetric and asymmetric deformation modes of ν$_4$ O-OP, while the intense absorption bands in the range of 1040–1090 cm^{-1} correspond to the ν$_3$ P-O [69–71].

The absorption bands from 3356 cm^{-1} marked with arrow refer to the bending oscillation modes of [OH]$^-$ structurally bound in the HA lattice that is present for Gen-Os but is overlapped with absorption bands characteristic of proteins or lipid functional groups, as well as physically adsorbed water. The absence of absorption bands for O-H is mentioned in different papers as a characteristic of biological apatite, explainable for xenografts such as Bio-Oss. In addition, for specimen HA1, the [OH] group absorption band in the range of 3600–3200 cm^{-1} is missing, which could prove that the hydroxyapatite synthesis from the natural source (egg-shells) leads to an analogous composition to bio-apatite from Bio-Oss. The presence of the [CO_3^{2-}] group in HA1 that substitutes [PO_4^{3-}] groups from the HA lattice, with absorption peaks reported at wave numbers of 1454 cm^{-1}, 1420 cm^{-1}, and 874 cm^{-1} as in natural bone, could be proof of compositional similarity. The strong absorption band at 874 cm^{-1} was assigned to the presence of CO_3^{2-} involved in B-type PO_4^{3-} substitution (ν$_{2[B]}$), and the absence of a lower intensity for the absorption band at 880 cm^{-1} attributed to A-type carbonate substitution (ν$_{2[A]}$ in OH$^-$ position) together with the missing band at around 3570 cm^{-1} for the stretching mode of OH$^-$ could be another piece of evidence. There are papers that presented methods to appreciate the content of CO_3^{2-} that substitutes PO_4^{3-} in bio-apatite (C/P), by measuring the amplitude of the absorption band at 1415–1420 cm^{-1} (as CO_3^{2-} quantity) and at 1011–1042 cm^{-1} corresponding to PO_4^{3-} quantity [69–71].

According to this method, the C/P ratio for HA1 was 0.028, that for BioOss was 0.125, and the highest content of carbonate for GenOs was 0.574, the latter containing a certain quantity of denatured collagen but without any confusion, and the characteristic absorption band for carbonyl (CO) group is placed at 1455 cm^{-1} [69–71].

Besides, absorption peaks at 2920 cm^{-1} (reported at 2923 cm^{-1}) and 2851 cm^{-1} are assigned to the stretching vibrations of CH, and wavenumbers of 1230, 1541, and 1648 cm^{-1} are absorption bands for C-H bonds in (CH2) and (CH3), and C-N, N-H, and C=O, respectively, for amide I, amide II, and amide III (reported absorption bands with maximum amplitude at wavenumbers of 1659–1555 cm^{-1}), and 1011 cm^{-1} (reported at 1035 and 1079 cm^{-1}) is assigned for bond vibrations ν(C–O) and ν(C–O–C), all of which have been reported for collagen type I [68].

The shape of the grains for the two xenografts resembles fragments of cancellous or cortical natural bone [18,62,63], while for the HA1 sample, the grains are very small and of prismatic shape (Figure 3g). At higher magnifications, the SEM images of the two xenografts show the characteristic morphology of the extracellular matrix in the bone tissue, with a very dense structure (Figure 3b,f), HA platelets crystallized in parallel planes, with nanometric width and 59–62 nm diameter, along with collagen fibers with hydroxyapatite mineralized on surface (Figure 3 c). For Gen-Os and Bio-Oss, the structure is caused by organic–inorganic nanocomposite biogenesis. Contrariwise, the HA1 sample SEM images indicate a soft and fluffy structure characteristic of the crystallization from precipitate, ultra-fine nanoparticle aggregate structures (5.44–7.69 nm), and no visible intergranular limits (Figure 3h,i) [74]. In a saturated solution, heterogeneous nucleation always takes place earlier than the homogeneous one, due to lower nucleus-free energy on foreign bodies [75].

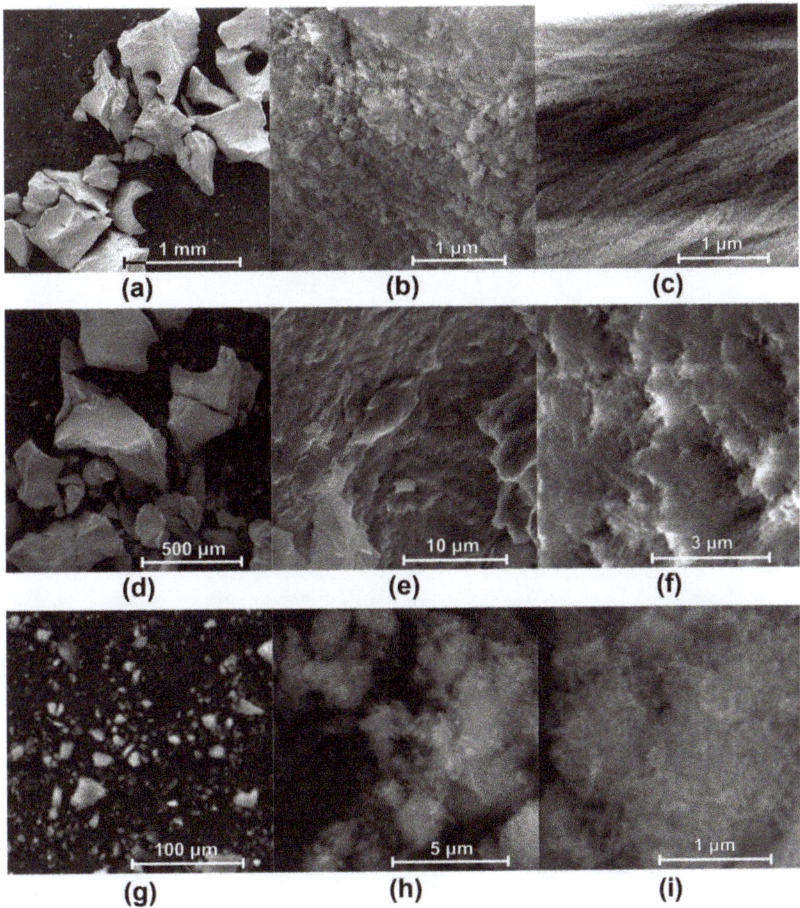

Figure 3. SEM images for: (**a–c**) Gen-Os (200×, 10,000×, and 40,000×); (**d–f**) Bio-Oss (200×, 10,000×, and 40,000×); (**g–i**) HA1 synthesized from egg-shells (1000×, 20,000×, and 100,000×).

The chemical elements identified in each sample are presented in the EDS spectra (Figure 4), and the elemental composition of HA1 and egg-shells before annealing powders can be observed in the XRF results (Table 3). The identified elements in the Bio-Oss and Gen-Os sample (C, P, O, and Ca) can be assumed to form a single phase of carbonated apatite with molar ratios of Ca:P~1.65 and 1.60, respectively. This is a well-known deviation from a hydroxyapatite unit cell stoichiometry (Ca/P = 1.67), which is in good correlation with the FTIR and XRD analysis results, where the nature of the sample source and the data is reported in the literature.

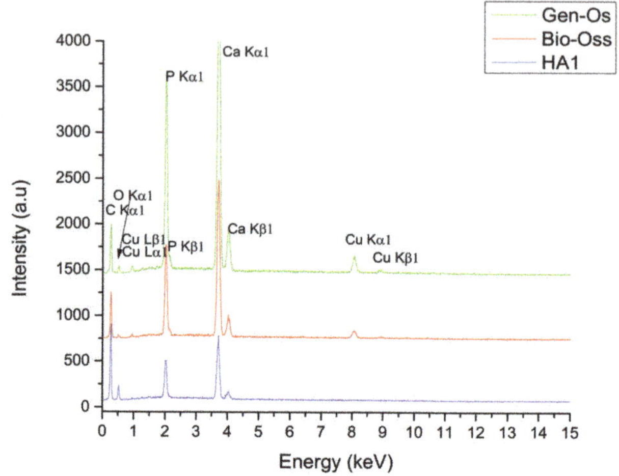

Figure 4. EDS spectra for: (red) Bio-Oss; (green) Gen-Os; (blue) HA1 powders.

Table 3. Main elemental constituents of HA1 powder and egg-shells before annealing by XRF.

Identified Element	Egg-Shell before Calcination		HA1	
	(% wt.)	Est. Error (%)	(% wt.)	Est. Error (%)
Ca	96.38	0.09	69.44	0.23
Na	1.82	0.07	0.627	0.23
Mg	0.980	0.049	0.594	0.044
P_x	0.334	0.017	29.14	0.030
S_x	0.233	0.012	0.0540	0.0027
K	0.0669	0.0033	-	-
Sr	0.0626	0.0031	0.0381	0.0019
Si	0.0420	0.0027	0.0400	0.0049
Al	0.0185	0.0055	-	-
Cl	0.0163	0.0009	0.0134	0.0022
Fe	0.0141	0.0016	0.0092	0.0027

Comparing the results of the elemental analyses from the EDS dispersion spectra, which have only a qualitative value, the presence of copper in the two xenograft samples is highlighted, in a higher content at the Gen-Os sample. In addition, an evident carbon content is observed in all three samples, which may arise from the carbonate groups in Bio-Oss and HA1, but may have an organic and inorganic nature in Gen-Os. The amplitude peak corresponding to the C element, found at around 0.3 eV, could be attributed to graphite conductive tape, too. None of the EDS spectra for any of the three samples mark the presence of trace elements, as reported in the literature [18,36–38].

Studies in the literature have reported biological apatite derived from bone products with a Ca/P ratio between 1.50 and 1.85, strongly dependent on bone species and the age factor [36]. The responsible factors for this stoichiometry deviation are the cationic and anionic substitutions of calcium, phosphate, or hydroxyl groups from the hydroxyapatite lattice with trace elements and carbonate or silicate groups, respectively [18]. Using the EDS spectrum for HA1 (Figure 4), the calculated ratio of Ca/P = 1.69 ± 0.1 exceeds the ratio of these two elements in stoichiometric hydroxyapatite, with the same ratio being reported in the literature for synthesized hydroxyapatite but also for biologic apatite provided from biogenic sources.

XRF analysis on the synthesized hydroxyapatite sample (HA1) and unannealed egg-shell powder used as a source for this synthesis (Table 3) showed the presence of important trace elements in the egg-shell source, kept during all treatments, and also found in a smaller amount in HA1 nano-powder. Hence, the Na content decreased from $1.82 \pm 0.07\%$ wt. in egg-shells to $0.63 \pm 0.23\%$ wt. in HA1, the Mg from $0.98 \pm 0.05\%$ to $0.59 \pm 0.04\%$, and so on. Such elements are known to be of great importance in the biocompatibility of natural hydroxyapatite and beneficial for the synthesized HA1. Many trace elements have been reported to substitute Ca ions of biologic HA, such as 0.9%wt Na^+ and 0.5%wt Mg^{2+}; $Sr^{2+} < 0.1\%$. PO_4^{3-} groups are also usually substituted by around 4–6% CO_3^{2-} [36]. In fact, bone mineral is a calcium-deficient apatite, where a Ca:P ratio of 1.67 is only the theoretical value for pure hydroxyapatite [36].

TEM images for Bio-Oss, Gen-Os, and HA1 samples can be seen in Figure 5A–H. At smaller magnification, crystalline aggregates composed of polyhedral and rod-like hydroxyapatite particles are observed for Bio-Oss and HA1 samples (Figure 5D orange arrows and Figure 5G), while for the Gen-Os sample, the rod-like particles are faded by the presence of remnant collagenous components. The rod-like morphology is confirmed at the higher magnification (Figure 5B,G), resembling both Bio-Oss and HA1 samples. However, a higher surface roughness and numerous internal pores can be observed in the HA1 sample (Figure 5C,H, yellow arrows). Figure 6A,B show Bio-Oss particle mean lengths and diameters of 44.6 nm and 8.3 nm, respectively. Larger particles were measured for HA1, with a mean diameter of 19.96 nm and mean length of 61.18 nm, and the crystallites were elongated in the [002] growth direction under the c axis, as also proven by XRD analysis results (Figure 1 right). The diffraction rings presented in the SAED patterns are similar for HA1, Bio-Oss, and Gen-Os (Figure 5C.1,E.1,H.1), for each crystalline plane identified, proving the polycrystalline hydroxyapatite as the main component of the three samples. The ring associated with the (0 0 2) plane is present in all SAED patterns, which is characteristic to natural bone due to the collagen fibrils, and is evidence of the compositional similarity between the three samples [36]. The (1 1 2), (2 1 1), and (3 0 0) planes form three rings that overlap for all samples, appearing brighter. The diffuse light in the SAED pattern marks the presence of an amorphous phase in Gen-OS.

For the Gen-Os, a sample of partially deproteinized porcine bone, the TEM images show the same polyhedral and nano-rod-like morphologies, with a mean length of 88.27 nm and mean diameter of 12.68 nm, larger than Bio-Oss. A possible explanation for these findings can be the reminiscence of natural bone proteins under the mineral structure, already proved by FTIR and SEM analysis. However, the sizes of Gen-Os particles are longer than those of HA1 (Figure 6A). Studies in the literature have reported bone hydroxyapatite crystals sizes to be 30–50 nm in length and 15–30 nm in width [36]. The width distribution of comparative particles is observed in Figure 6 B. In addition, the thinnest hydroxyapatite rod-like particles are contained by the Bio-Oss specimen (Figure 6B red), with the measured particle diameters for HA1 being of comparable size as those of Gen-Os.

The obtained results are in good agreement with previous literature studies, which have associated the morphology of synthetic HA with the hydrothermal conditions [51,52]. Hence, when treated at 150 °C for 24 h, needle-like structures are obtained [55], while for 72 h aging, the observed morphology is rod-like [51,52,59]. As could be seen in TEM, the micrographs associated with the HA1 sample (Figure 5G,H), the nano-rod-like crystals have high rough surfaces created by hydrothermal MW-assisted treatment of the HA precipitate without any templates. The rounded pores are irregularly distributed (Figure 5H, yellow arrow) and present on both the particle's surface, as well as between grains, having an average size of 3.3 nm (Figure 6C) [60]. Comparing the TEM image of Bio-Oss in Figure 5C with the HA1 sample (Figure 5H), an obvious similar roughness of the particle's surfaces could be seen, even if their synthesis history is different. Nevertheless, the sample HA1 has a very high internal porosity compared with Bio-Oss (Figure 5C,H yellow arrow).

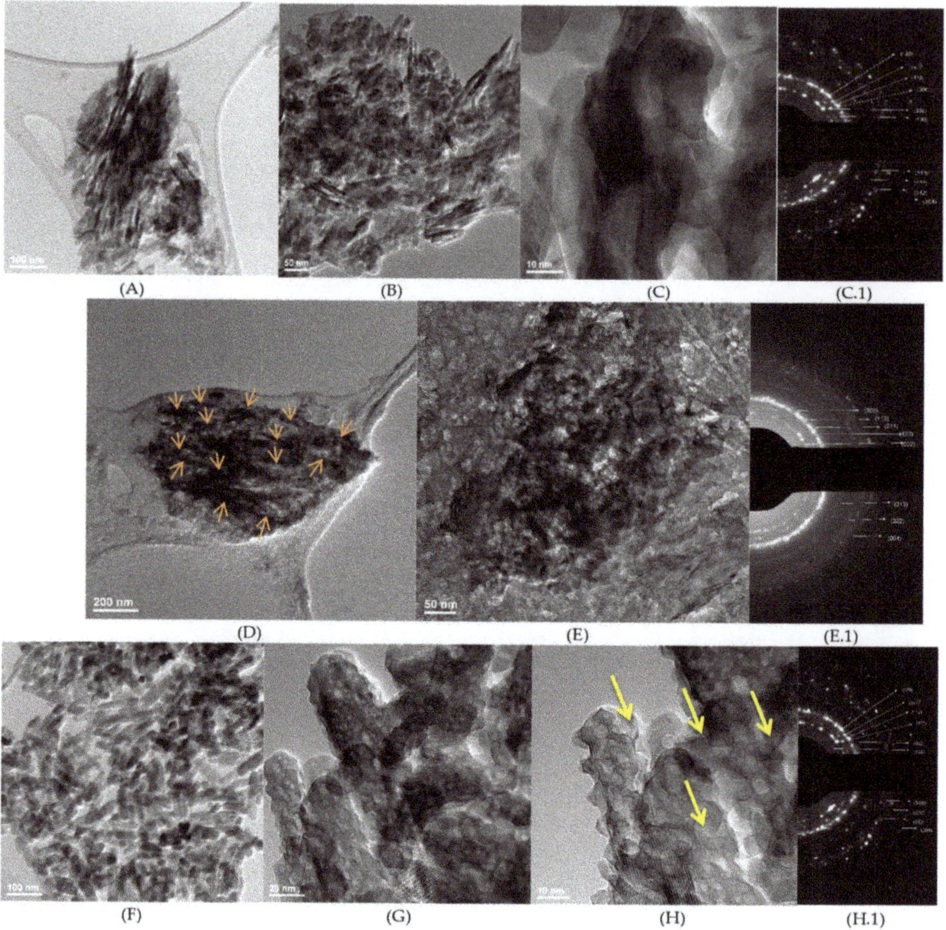

Figure 5. TEM images and SAED patterns for: (**A–C**) Bio-Oss, (**C.1**) SAED pattern Bio-Oss; (**D,E**) Gen-Os (rod-like particles-red arrows), (**E.1**) SAED pattern Gen-Os; (**F–H**) HA1 (intra-particle pores-yellow arrows), (**H.1**) SAED pattern HA1.

High-resolution transmission electron microscopy (HRTEM) images for every three specimens are presented in Figure 7a–c. This investigation allows the measurement of crystallites sizes on the selected area, as well as determining the *d*-spacing of the family of parallel planes belonging to a crystallite oriented under a certain direction in the polycrystalline particle, by measuring the distance between crystallite atom planes. In Figure 7a (inset), the d-spacing of 5.2881 Å corresponds to (101) Miller's indices of hydroxyapatite crystals in the Bio-Oss sample, 2.7131 Å is the d-spacing for parallel planes in direction (300) in Gen-Os (Figure 7b), while the crystalline plane (112) has a *d*-spacing of 2.7613 Å for HA1 (Figure 7c). Some differences between the crystallite mean size determined from HRTEM and those calculated by Rietveld equations with XRD data were observed. A possible explanation could be the fact that HRTEM images show only part of a particle, with few crystallites grown under some crystalline planes, while the Rietveld method provides the mean size of crystallites, taking into consideration all crystalline planes of the sample.

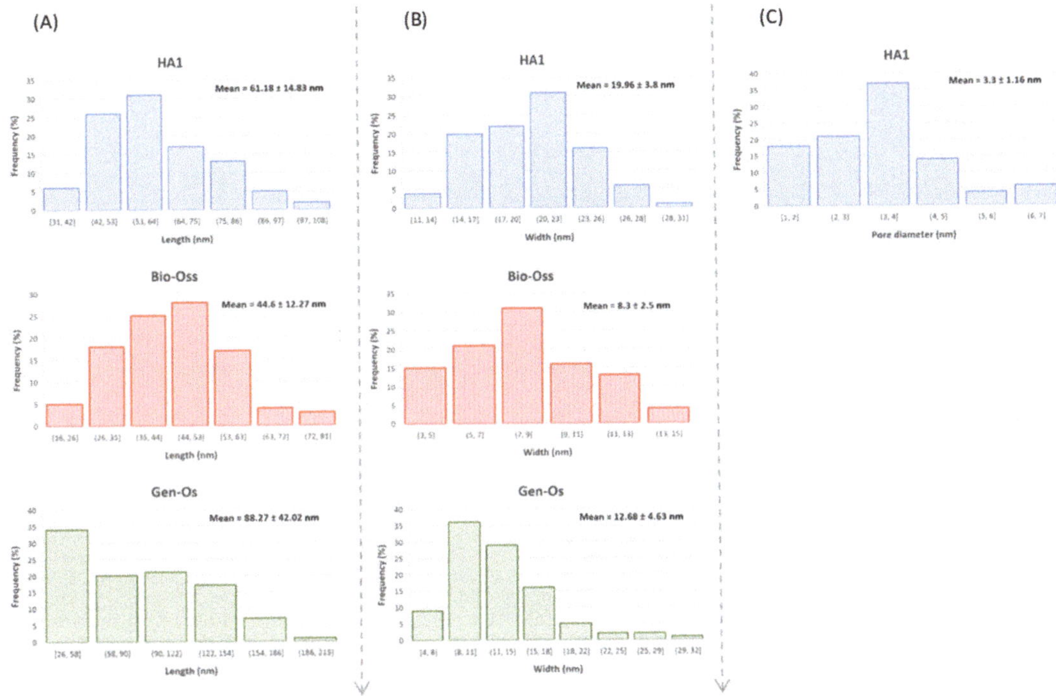

Figure 6. HA1 (blue), Bio-Oss (red), Gen-Os (green) particles length distribution (**A**) and width distribution (**B**); internal pores size distribution for HA1 sample (**C**).

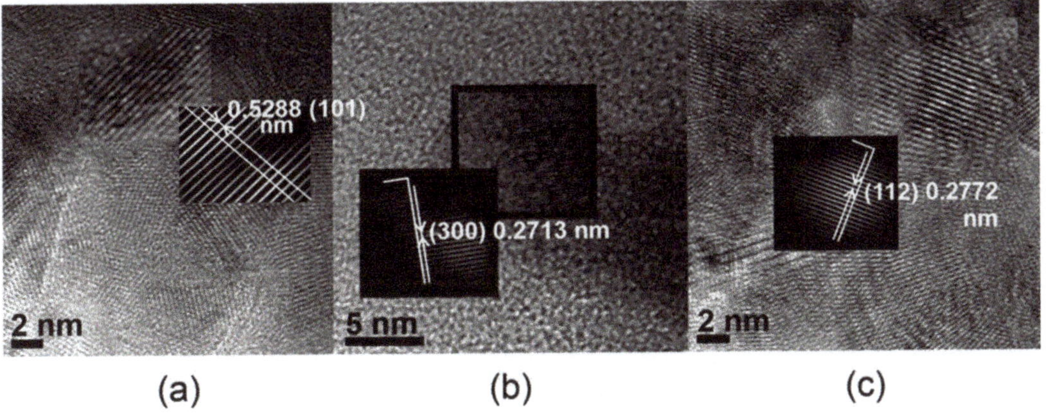

Figure 7. HRTEM images for specimens: (**a**) Bio-Oss (inset, inverse fast Fourier transform); (**b**) Gen-Os (inset, inverse fast Fourier transform); (**c**) HA1 (inset, inverse fast Fourier transform).

The percentage of viable cells after the 72 h incubation period (Figure 8A) was calculated by taking the ratio between the absorbance recorded on cell cultures in the presence of biomaterial powders and that of the control sample (CTRL) [76–78]. The metabolic activity of AFSCs in the presence of HA1 powder proves to be higher than the cell viability in the presence of the two xenografts, with the lowest viability in the series being registered in the presence of Bio-Oss powder. However, the one-way analysis of variance (ANOVA), followed by a two-tailed *t*-test with Bonferroni post-hoc correction results, showed that the differences between Bio-Oss, Gen-Os, and HA1 samples are not statistically significant. Yet, there is a statistically significant difference between the cellular viability registered in their presence and the control sample, which contains only AFSCs. Hence, even though the analyzed powders are inhibited in a small proportion, the metabolic activity of the AFSCs, HA1 sample proves similar enough with the two commercial xenografts already used as a bone substitute in dentistry. Consequently, the already mentioned morphological and structural differences between Bio-Oss, Gen-Os, and HA1 do not greatly influence their biocompatibility.

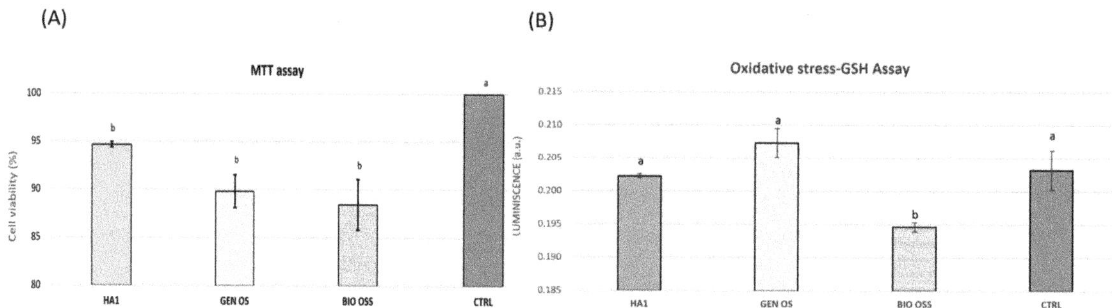

Figure 8. (**A**) MTT assay results (after 72 h incubation) and (**B**) GSH assay showing the oxidative stress of AFSCs cultured in the presence of HA1, Gen-Os, Bio-Oss powders, and CTRL sample (only cells); the results are presented as the mean ± S.D. of 3 replicates; different letters indicate significant differences between each sample; $p < 0.05/n$ (n = 6)-based ANOVA statistical analysis, followed by a two-tailed *t*-test with Bonferroni post hoc correction (**A**).

In the GSH assay, the cytotoxic effect translates into an increased oxidative stress generated on the cells in the presence of HA1, Gen-Os, or Bio-Oss samples, and for the control sample, consisting of cells only. An intense AFSC bioactivity leads to more GSH enzymes coupled to fight against oxidative species, forming glutathione disulfide (GSSG). The GSH/GSSG molar ratio represents a powerful index of oxidative stress and disease risk and can be determined by numerous analytical methods, including UV-Vis spectrophotometry. The measured luminescence is proportional to the amount of GSH involved in the antioxidative stress, and as shown in Figure 8B, the most stress-free AFSCs are those in contact with Bio-Oss, with the HA1 sample coming next. The CTRL sample is placed before Gen-Os, which manifests the strongest oxidative stress on the cell line cultured, all incubated for 24 h. However, data analysis results showed that the differences between CTRL, Gen-Os, and HA1 samples are not statistically significant. Yet, there is a statistically significant difference between the oxidative stress registered in their presence and the Bio-Oss sample. Even though the Gen-Os and HA1 powders seem to cause an increase in GSH amount, unlike Bio-Oss, their cytotoxic effect is nonsignificant compared to cells only (CTRL) and can be considered biocompatible, similar to MTT analysis results. The GSH test proves once more the biological resemblance between the HA1 sample and Gen-Os xenograft, but also highlights the atypical behavior of Bio-Oss. Analyzing the morphological and structural characteristics of Bio-Oss, compared with Gen-Os and HA1, a prospective correlation with their biological properties arises: smaller particles manage to diminish the oxidative stress level by restoring the balance between the formation of reactive oxygen species (ROS) in cells and the capability of the cells to clear these free radicals; the smallest hydroxyapatite rod-like particles were observed for the Bio-Oss sample, while the measured particles for HA1 were comparable in size with Gen-Os.

The fluorometric microculture cytotoxicity assay (FMCA) is an in vitro nonclonogenic-based cell viability assessment used for the cytotoxic and cytostatic measurement effect of different compounds or biomaterials, after a short time of incubation. The assay is based on fluorescein diacetate (FDA) hydrolysis by esterase in cells, keeping intact plasma membranes [79]. AFSCs were seeded under the described protocol (control cells only) and in contact with Bio-Oss, Gen-Oss, and HA1 samples, which could be observed as viable at the fluorescence microscope, after 72 h of incubation, with cells absorbing CMTPX fluorophores added in the cytoplasm (Figure 9). Fluorescence microscopy images show that AFSCs are viable and preserve their initial morphology, with homogenous sizes and density distributions in the culture well plates. No fragmented or dead cells could be identified on the background in the presence of xenografts and HA1 powders, as well as without any biomaterials in contact. In addition, from the images, the formation of filopodia actin-rich protrusions attached by many cells could be seen. As they are usually involved in numerous cellular processes, including cell migration, wound healing, adhesion to the extracellular matrix, and guidance toward chemoattractants, it demonstrates that seeded AFSCs exhibit a good comparable bioactivity for all samples analyzed [79].

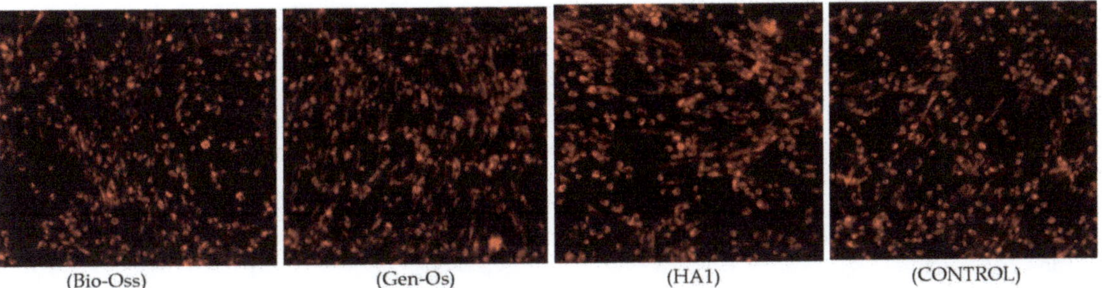

Figure 9. Fluorescent microscopy images of Bio-Oss, Gen-Os, HA1, and CONTROL colored with CMTPX fluorophore.

4. Conclusions

Nano-hydroxyapatite (HA1) was synthesized from egg-shells by the microwave-assisted hydrothermal method (HTMW). From this hybrid, through an unconventional genesis route, a crystalline, homogenous in dimension, rod-like hydroxyapatite was obtained, after only several minutes, compared to various literature studies that reported several hours when using a hydrothermal technique alone. The obtained material demonstrates a mimetic composition, morphology, and structure with the commercial xenografts Bio-Oss® and Gen-OS®. The fact that the HA1 sample, unlike the two xenografts, proved to have a very high meso-porosity was noticeable. This could be associated with an improved biomolecule adhesion and a potential increased osteoconductivity, and could be the cause for the good results of this sample at all in vitro cytotoxicity assays. Moreover, HA1 can be utilized in granular form, as well as xenografts, with better bioactivity and osteoconductivity than the hydroxyapatite-based scaffolds. The HA1 powder thus synthesized has a high potential for applications in bone substitution, teeth fillers, or drug delivery systems.

Author Contributions: Conceptualization, C.R.D., I.A.N. and E.A.; funding acquisition, A.F. and E.A.; investigation, V.A.S., A.I.N., F.I. and R.T.; methodology, C.R.D. and I.A.N.; resources, L.T.C.; supervision, A.F. and E.A.; visualization, V.A.S. and R.T.; writing—original draft, C.R.D.; writing—review and editing, I.A.N. and V.A.S. All authors have read and agreed to the published version of the manuscript.

Funding: This research was funded by UEFISCDI project "Innovative biomaterials for treatment and diagnosis," BIONANOINOV grant number PN-IIIP1-1.2-PCCD-I2017-0629 and the project *ANTREPRENORDOC*, in the framework of the Human Resources Development Operational Programme 2014–2020, financed from the European Social Fund under the contract number 36355/23.05.2019 HRD OP /380/6/13—SMIS Code: 123847.

Data Availability Statement: Not applicable.

Acknowledgments: The support of the EU-funding project POSCCE-A2-O2.2.1-2013-1/Priority Axe 2, Project No. 638/12.03.2014, ID 1970, SMIS-CSNR code 48652 is gratefully acknowledged for the equipment purchased from this project.

References

1. Castano, I.M.; Curtin, C.M.; Duffy, G.P.; O'Brien, F.J. Next generation bone tissue engineering: Non-viral miR-133a inhibition using collagen-nanohydroxyapatite scaffolds rapidly enhances osteogenesis. *Sci. Rep.* **2016**, *6*, 27941. [CrossRef] [PubMed]
2. Sohn, H.S.; Oh, J.K. Review of bone graft and bone substitutes with an emphasis on fracture surgeries. *Biomater. Res.* **2019**, *23*, 9. [CrossRef]
3. Mello, B.F.; de Carvalho Formiga, M.; de Souza da Silva, L.F.; Dos Santos Coura, G.; Shibli, J.A. Horizontal ridge augmentation using a xenograft bone substitute for implant-supported fixed rehabilitation: A case report with four years of follow-up. *Case Rep. Dent.* **2020**, *2020*, 6723936. [CrossRef] [PubMed]
4. Iviglia, G.; Kargozar, S.; Baino, F. Biomaterials, Current Strategies, and Novel Nano-Technological Approaches for Periodontal Regeneration. *J. Funct. Biomater.* **2019**, *10*, 3. [CrossRef]
5. Ionescu, O.A.; Ciocilteu, M.; Manda, V.; Neacsu, I.; Ficai, A.; Amzoiu, E.M.; Turcu-Stiolica, A.; Croitoru, O.; Neamtu, J.O. Bone—Graft delivery systems of type PLGA-gentamicin and Collagen—hydroxyapatite—gentamicine. *Mater. Plast.* **2019**, *56*, 534–537. Available online: http://www.revmaterialeplastice.ro (accessed on 15 May 2021). [CrossRef]
6. Haugen, H.J.; Lyngstadaas, S.P.; Rossi, F.; Perale, G. Bone grafts: Which is the ideal biomaterial? *J. Clin. Periodontol.* **2019**, *46* (Suppl. 21), 92–102. [CrossRef] [PubMed]
7. Titsinides, S.; Agrogiannis, G.; Karatzas, T. Bone grafting materials in dentoalveolar reconstruction: A comprehensive review. *Jpn. Dent. Sci. Rev.* **2019**, *55*, 26–32. [CrossRef]
8. Zhou, Y.; Wu, C.; Chang, J. Bioceramics to regulate stem cells and their microenvironment for tissue regeneration. *Mater. Today* **2019**, *24*, 41–56. [CrossRef]
9. Pajarinen, J.; Lin, T.H.; Nabeshima, A.; Jämsen, E.; Lu, L.; Nathan, K.; Yao, Z.; Goodman, S.B. Mesenchymal stem cells in the aseptic loosening of total joint replacements. *J. Biomed. Mater. Res. Part A* **2017**, *105*, 1195–1207. [CrossRef] [PubMed]
10. Pu'ad, N.A.S.M.; Koshy, P.; Abdullah, H.Z.; Idris, M.I.; Lee, T.C. Syntheses of hydroxyapatite from natural sources. *Heliyon* **2019**, *5*, e01588. [CrossRef]
11. Lertcumfu, N. Properties of calcium phosphates ceramic composites derived from natural materials. *Ceram. Int.* **2016**, *42*, 10638–10644. [CrossRef]
12. Ofudje, E.A.; Rajendran, A.; Adeogun, A.I.; Idowu, M.A.; Kareem, S.O.; Pattanayak, D.K. Synthesis of organic derived hydroxyapatite scaffold from pig bone waste for tissue engineering applications. *Adv. Powder Technol.* **2018**, *29*, 1–8. [CrossRef]
13. Janus, A.M.; Faryna, M.; Haberko, K.; Rakowska, A.; Panz, T. Chemical and microstructural characterization of natural hydroxyapatite derived from pig bones. *Microchim. Acta* **2007**, *161*, 349–353. [CrossRef]
14. Khanijou, M.; Seriwatanachai, D.; Boonsiriseth, K.; Suphangul, S.; Pairuchvej, V.; Srisatjaluk, R.L.; Wongsirichat, N. Bone graft material derived from extracted tooth: A review literature. *J. Oral Maxillofac. Surg. Med. Pathol.* **2019**, *31*, 1–7. [CrossRef]
15. Mijiritsky, E.; Ferroni, L.; Gardin, C.; Bressan, E.; Zanette, G.; Piattelli, A.; Zavan, B. Porcine bone scaffolds adsorb growth factors secreted by MSCs and improve bone tissue repair. *Materials* **2017**, *10*, 1054. [CrossRef]
16. Londoño-Restrepo, S.M.; Ramirez-Gutierrez, C.F.; del Real, A.; Rubio-Rosas, E.; Rodriguez-García, M.E. Study of bovine hydroxyapatite obtained by calcination at low heating rates and cooled in furnace air. *J. Mater. Sci.* **2016**, *51*, 4431–4441. [CrossRef]
17. Figueiredo, M.; Henriques, J.; Martins, G.; Guerra, F.; Judas, F.; Figueiredo, H. Physicochemical characterization of biomaterials commonly used in dentistry as bone substitutes—Comparison with human bone. *J. Biomed. Mater. Res. B Appl. Biomater.* **2010**, *92*, 409–419. [CrossRef]
18. Cassetta, M.; Perrotti, V.; Calasso, S.; Piattelli, A.; Sinjari, B.; Iezzi, G. Bone formation in sinus augmentation procedures using autologous bone, porcine bone, and a 50:50 mixture: A human clinical and histological evaluation at 2 months. *Clin. Oral Implant. Res.* **2015**, *26*, 1180–1184. [CrossRef]
19. Sun, R.-X.; Lv, Y.; Niu, Y.-R.; Zhao, X.-H.; Cao, D.-S.; Tang, J.; Sun, X.-C.; Chen, K.-Z. Physicochemical and biological properties of bovine-derived porous hydroxyapatite/collagen composite and its hydroxyapatite powders. *Ceram. Int.* **2017**, *43*, 16792–16798. [CrossRef]
20. Sadat-Shojai, M.; Khorasani, M.-T.; Dinpanah-Khoshdargi, E.; Jamshidi, A. Synthesis methods for nanosized hydroxyapatite with diverse structures. *Acta Biomater.* **2013**, *9*, 7591–7621. [CrossRef]
21. Sunil, B.R.; Jagannatham, M. Producing hydroxyapatite from fish bones by heat treatment. *Mater. Lett.* **2016**, *185*, 411–414. [CrossRef]
22. Pon-On, W.; Suntornsaratoon, P.; Charoenphandhu, N.; Thongbunchoo, J.; Krishnamra, N.; Tang, I.M. Hydroxyapatite from fish scale for potential use as bone scaffold or regenerative material. *Mater. Sci. Eng. C Mater. Biol. Appl.* **2016**, *62*, 183–189. [CrossRef]

23. Goh, K.W.; Wong, Y.H.; Ramesh, S.; Chandran, H.; Krishnasamy, S.; Ramesh, S.; Sidhu, A.; Teng, W.D. Effect of pH on the properties of eggshell-derived hydroxyapatite bioceramic synthesized by wet chemical method assisted by microwave irradiation. *Ceram. Int.* **2021**, *47*, 8879–8887. [CrossRef]

24. Mohd Pu'ad, N.A.S.; Alipal, J.; Abdullah, H.Z.; Idris, M.I.; Lee, T.C. Synthesis of eggshell derived hydroxyapatite via chemical precipitation and calcination method. *Mater. Today Proc.* **2021**, *42*, 172–177. [CrossRef]

25. Cestari, F.; Agostinacchio, F.; Galotta, A.; Chemello, G.; Motta, A.; Sglavo, V.M. Nano-Hydroxyapatite Derived from Biogenic and Bioinspired Calcium Carbonates: Synthesis and In Vitro Bioactivity. *Nanomaterials* **2021**, *11*, 264. [CrossRef] [PubMed]

26. Núñez, D.; Elgueta, E.; Varaprasad, K.; Oyarzún, P. Hydroxyapatite nanocrystals synthesized from calcium rich bio-wastes. *Mater. Lett.* **2018**, *230*, 64–68. [CrossRef]

27. Suresh Kumar, C.; Dhanaraj, K.; Vimalathithan, R.M.; Ilaiyaraja, P.; Suresh, G. Hydroxyapatite for bone related applications derived from sea shell waste by simpleprecipitation method. *J. Asian Ceram. Soc.* **2020**, *8*, 416–429. [CrossRef]

28. Pal, A.; Maity, S.; Chabri, S.; Bera, S.; Chowdhury, A.R.; Das, M.; Sinha, A. Mechanochemical synthesis of nanocrystalline hydroxyapatite from Mercenaria clam shells and phosphoric acid. *Biomed. Phys. Eng. Express* **2017**, *3*, 015010. [CrossRef]

29. Li, Y.; Chen, S.K.; Li, L.; Qin, L.; Wang, X.L.; Lai, Y.X. Bone defect animal models for testing efficacy of bone substitute biomaterials. *J. Orthop. Transl.* **2015**, *3*, 95–104. [CrossRef] [PubMed]

30. Yamada, M.; Egusa, H. Current bone substitutes for implant dentistry. *J. Prosthodont. Res.* **2018**, *62*, 152–161. [CrossRef]

31. Prasadh, S.; Wong, R.C.W. Unraveling the mechanical strength of biomaterials used as a bone scaffold in oral and maxillofacial defects. *Oral Sci. Int.* **2018**, *15*, 48–55. [CrossRef]

32. Przekora, A. Current trends in fabrication of biomaterials for bone and cartilage regeneration: Materials modifications and biophysical stimulations. *Int. J. Mol. Sci.* **2019**, *20*, 435. [CrossRef]

33. Pajarinen, J.; Lin, T.; Gibon, E.; Kohno, Y.; Maruyama, M.; Nathan, K.; Lu, L.; Yao, Z.; Goodman, S.B. Mesenchymal stem cell-macrophage crosstalk and bone healing. *Biomaterials* **2019**, *196*, 80–89. [CrossRef] [PubMed]

34. Selders, G.S.; Fetz, A.E.; Radic, M.Z.; Bowlin, G.L. An overview of the role of neutrophils in innate immunity, inflammation and host-biomaterial integration. *Regen Biomater* **2017**, *4*, 55–68. [CrossRef]

35. Dewi, A.H.; Ana, I.D. The use of hydroxyapatite bone substitute grafting for alveolar ridge preservation, sinus augmentation, and periodontal bone defect: A systematic review. *Heliyon* **2018**, *4*, e00884. [CrossRef]

36. Olszta, M.J.; Cheng, X.; Jee, S.S.; Kumar, R.; Kim, Y.-Y.; Kaufman, M.J.; Douglas, E.P.; Gower, L.B. Bone structure and formation: A new perspective. *Mater. Sci. Eng. R Rep.* **2007**, *58*, 77–116. [CrossRef]

37. Erdem, U.; Dogan, M.; Metin, A.U.; Baglar, S.; Turkoz, M.B.; Turk, M.; Nezir, S. Hydroxyapatite-based nanoparticles as a coating material for the dentine surface: An antibacterial and toxicological effect. *Ceram. Int.* **2020**, *46*, 270–280. [CrossRef]

38. Agbeboh, N.I.; Oladele, I.O.; Daramola, O.O.; Adediran, A.A.; Olasukanmi, O.O.; Tanimola, M.O. Environmentally sustainable processes for the synthesis of hydroxyapatite. *Heliyon* **2020**, *6*, e03765. [CrossRef]

39. Koon, H.E.C.; Nicholson, R.A.; Collins, M.J. A practical approach to the identification of low temperature heated bone using TEM. *J. Archaeol. Sci.* **2003**, *30*, 1393–1399. [CrossRef]

40. Santhosh, S.; Balasivanandha Prabu, S. Thermal stability of nano hydroxyapatite synthesized from sea shells through wet chemical synthesis. *Mater. Lett.* **2013**, *97*, 121–124. [CrossRef]

41. Salma-Ancane, K.; Stipniece, L.; Irbe, Z. Effect of biogenic and synthetic starting materials on the structure of hydroxyapatite bioceramics. *Ceram. Int.* **2016**, *42*, 9504–9510. [CrossRef]

42. Morgan, H.; Willson, R.M.; Elliott, J.C.; Dowker, S.E.P.; Anderson, P. Preparation and characterisation of monoclinic hydroxyapatite and its precipitated carbonate apatite intermediate. *Biomaterials* **2000**, *21*, 617–627. [CrossRef]

43. Singh, G.; Singh, R.P.; Jolly, S.S. Customized hydroxyapatites for bone-tissue engineering and drug delivery applications: A review. *J. Sol. Gel Sci. Technol.* **2020**, *94*, 505–530. [CrossRef]

44. Tabaght, F.E.; Azzaoui, K.; Elidrissi, A.; Hamed, O.; Mejdoubi, E.; Jodeh, S.; Akartasse, N.; Lakrat, M.; Lamham, A. New nanostructure based on hydroxyapatite modified cellulose for bone substitute, synthesis, and characterization. *Int. J. Polym. Mater. Polym. Biomater.* **2020**. [CrossRef]

45. Zou, Z.; Lin, K.; Chen, L.; Chang, J. Ultrafast synthesis and characterization of carbonated hydroxyapatite nanopowders via sonochemistry-assisted microwave process. *Ultrason. Sonochem.* **2012**, *19*, 1174–1179. [CrossRef] [PubMed]

46. Noviyanti, A.R.; Akbar, N.; Deawati, Y.; Ernawati, E.E.; Malik, Y.T.; Fauzia, R.P.; Risdiana. A novel hydrothermal synthesis of nanohydroxyapatite from eggshell-calcium-oxide precursors. *Heliyon* **2020**, *6*, e03655. [CrossRef]

47. Moreno-Perez, B.; Matamoros-Veloza, Z.; Rendon-Angeles, J.C.; Yanagisawa, K.; Onda, A.; Pérez-Terrazas, J.E.; Mejia-Martínez, E.E.; Burciaga Díaz, O.; Rodríguez-Reyes, M. Synthesis of silicon-substituted hydroxyapatite using hydrothermal process. *Bol. Soc. Española Cerám. Y Vidr.* **2020**, *59*, 50–64. [CrossRef]

48. Nosrati, H.; Sarraf-Mamoory, R.; Le, D.Q.S.; Perez, M.C.; Buünger, C.E. Evaluation of argon-gas-injected solvothermal synthesis of hydroxyapatite crystals followed by high-frequency induction heat sintering. *Cryst. Growth Des.* **2020**, *20*, 3182–3189. [CrossRef]

49. Kharisov, B.I.; Kharissova, O.V.; Méndez, U.O. Microwave hydrothermal and solvothermal processing of materials and compounds. *Dev. Appl. Microw. Heat.* **2012**.

50. Karunakaran, G.; Cho, E.; Kumar, G.S.; Kolesnikov, E.; Janarthanan, G.; Pillai, M.M.; Rajendran, S.; Boobalan, S.; Gorshenkov, M.V.; Kuznetsov, D. Ascorbic acid-assisted microwave synthesis of mesoporous ag-doped hydroxyapatite nanorods from biowaste seashells for implant applications. *ACS Appl. Bio. Mater.* **2019**, *2*, 2280–2293. [CrossRef]

51. Razali, N.A.I.M.; Pramanik, S.; Osmana, N.A.A.; Radzib, Z.; Murphya, B.P. Conversion of calcite from cockle shells to bioactive nanorod hydroxyapatite for biomedical applications. *J. Ceram. Process. Res.* **2016**, *17*, 699–706.

52. Su, Y.; Wang, J.; Li, S.; Zhu, J.; Liu, W.; Zhang, Z. Self-templated microwave-assisted hydrothermal synthesis of two-dimensional holey hydroxyapatite nanosheets for efficient heavy metal removal. *Environ. Sci. Pollut. Res. Int.* **2019**, *26*, 30076–30086. [CrossRef]

53. Qi, Y.; Shen, J.; Jiang, Q.; Jin, B.; Chen, J.; Zhang, X. The morphology control of hydroxyapatite microsphere at high pH values by hydrothermal method. *Adv. Powder Technol.* **2015**, *26*, 1041–1046. [CrossRef]

54. Yu, H.-P.; Zhu, Y.-J.; Lu, B.-Q. Highly efficient and environmentally friendly microwave-assisted hydrothermal rapid synthesis of ultralong hydroxyapatite nanowires. *Ceram. Int.* **2018**, *44*, 12352–12356. [CrossRef]

55. Méndez-Lozano, N.; Velázquez-Castillo, R.; Rivera-Muñoz, E.M.; Bucio-Galindo, L.; Mondragón-Galicia, G.; Manzano-Ramírez, A.; Ocampo, M.Á.; Apátiga-Castro, L.M. Crystal growth and structural analysis of hydroxyapatite nanofibers synthesized by the hydrothermal microwave-assisted method. *Ceram. Int.* **2017**, *43*, 451–457. [CrossRef]

56. Castro, M.A.M.; Portela, T.O.; Correa, G.S.; Oliveira, M.M.; Rangel, J.H.G.; Rodrigues, S.F.; Mercury, J.M.R. Synthesis of hydroxyapatite by hydrothermal and microwave irradiation methods from biogenic calcium source varying pH and synthesis time. *Bol. Soc. Española Cerám. Vidrio* **2020**. [CrossRef]

57. Mondal, S.; Hoang, G.; Manivasagan, P.; Moorthy, M.S.; Phan, T.T.V.; Kim, H.H.; Nguyen, T.P.; Oh, J. Rapid microwave-assisted synthesis of gold loaded hydroxyapatite collagen nano-bio materials for drug delivery and tissue engineering application. *Ceram. Int.* **2019**, *45*, 2977–2988. [CrossRef]

58. Chen, J.; Liu, J.; Deng, H.; Yao, S.; Wang, Y. Regulatory synthesis and characterization of hydroxyapatite nanocrystals by a microwave-assisted hydrothermal method. *Ceram. Int.* **2020**, *46*, 2185–2193. [CrossRef]

59. Karunakaran, G.; Kumar, G.S.; Cho, E.-B.; Sunwoo, Y.; Kolesnikov, E.; Kuznetsov, D. Microwave-assisted hydrothermal synthesis of mesoporous carbonated hydroxyapatite with tunable nanoscale characteristics for biomedical applications. *Ceram. Int.* **2019**, *45*, 970–977. [CrossRef]

60. Hassan, M.N.; Mahmoud, M.M.; E-Fattaha, A.A.; Kandil, S. Microwave-assisted preparation of Nano-hydroxyapatite for bone substitutes. *Ceram. Int.* **2016**, *42*, 3725–3744. [CrossRef]

61. Stroe, S.; Naicu, V.; Patrascu, I. Composite materials based on fibrin and their role in regeneration of the periodontal osseos defect. *Rom. J. Mater.* **2014**, *44*, 189–194.

62. Stroe, S.; Vasilescu, V.; Calin, C.; Patrascu, I. SEM and EDX evaluations on biomineralization of newly-formed bone tissue. *Rom. J. Mater.* **2014**, *44*, 195–203.

63. Landi, E.; Tampieri, A.; Celotti, G.; Sprio, S. Densification behaviour and mechanisms of synthetic hydroxyapatites. *J. Eur. Ceram. Soc.* **2000**, *20*, 2377–2387. [CrossRef]

64. Rietveld, H.M. A profile refinement method for nuclear and magnetic structures. *J. Appl. Cryst.* **1969**, *2*, 65. [CrossRef]

65. Uvarov, V.; Popov, I. Metrological characterization of X-ray diffraction methods at different acquisition geometries for determination of crystallite size in nano-scale materials. *Mater. Charact.* **2013**, *85*, 111–123. [CrossRef]

66. Ahmadipour, M.; Abu, M.J.; Ab Rahman, M.F.; Ain, M.F.; Ahmad, Z.A. Assessment of crystallite size and strain of CaCu3Ti4O12 prepared via conventional solid-state reaction. *Micro Nano Lett.* **2016**, *11*, 147–150. [CrossRef]

67. Belbachir, K.; Noreen, R.; Gouspillou, G.; Petibois, C. Collagen types analysis and differentiation by FTIR spectroscopy. *Anal. Bioanal. Chem.* **2009**, *395*, 829–837. [CrossRef]

68. Antonakos, A.; Liarokapis, E.; Leventouri, T. Micro-Raman and FTIR studies of synthetic and natural apatites. *Biomaterials* **2007**, *28*, 3043–3054. [CrossRef]

69. Nagy, G.; Lorand, T.; Patonai, Z.; Montsko, G.; Bajnoczky, I.; Marcsik, A.; Mark, L. Analysis of pathological and non-pathological human skeletal remains by FT-IR spectroscopy. *Forensic Sci. Int.* **2008**, *175*, 55–60. [CrossRef] [PubMed]

70. Thompson, T.J.U.; Gauthier, M.; Islam, M. The application of a new method of Fourier Transform Infrared Spectroscopy to the analysis of burned bone. *J. Archaeol. Sci.* **2009**, *36*, 910–914. [CrossRef]

71. Airini, R.; Iordache, F.; Alexandru, D.; Savu, L.; Epureanu, F.B.; Mihailescu, D.; Amuzescu, B.; Maniu, H. Senescence-induced immunophenotype, gene expression and electrophysiology changes in human amniocytes. *J. Cell Mol. Med.* **2019**, *23*, 7233–7245. [CrossRef] [PubMed]

72. Iordache, F.; Constantinescu, A.; Andrei, E.; Amuzescu, B.; Halitzchi, F.; Savu, L.; Maniu, H. Electrophysiology, immunophenotype, and gene expression characterization of senescent and cryopreserved human amniotic fluid stem cells. *J. Physiol. Sci.* **2016**, *66*, 463–476. [CrossRef]

73. Barakat, N.A.M.; Khalil, K.A.; Sheikh, F.A.; Omran, A.M.; Gaihre, B.; Khil, S.M.; Kim, H.Y. Physiochemical characterizations of hydroxyapatite extracted from bovine bones by three different methods: Extraction of biologically desirable HAp. *Mater. Sci. Eng. C* **2008**, *28*, 1381–1387. [CrossRef]

74. Mosina, M.; Locs, J. Synthesis of Amorphous Calcium Phosphate: A Review. *Key Eng. Mater.* **2020**, *850*, 199–206. [CrossRef]

75. Morgan, D.M. *Tetrazolium (MTT) Assay for Cellular Viability and Activity*; Methods Mol. Biol.; Morgan, D., Ed.; Humana Press Inc.: Totowa, NJ, USA, 1998; Volume 79, pp. 179–183.

76. Van Meerloo, J.; Kaspers, G.J.; Cloos, J. Cell sensitivity assays: The MTT assay. In *Cancer Cell Culture: Methods and Protocols*, 2nd ed.; Methods in Molecular Biology; Cree, I.A., Ed.; Springer Science+Business Media: Berlin, Germany; Volume 731, pp. 237–245. [CrossRef]

77. Gola, J. Quality Control of Biomaterials—Overview of the Relevant Technologies. In *Stem Cells and Biomaterials for Regenerative Medicine*; Los, M.J., Hudecki, A., Wiechec, E., Eds.; Academic Press: Cambridge, MA, USA, 2019. [CrossRef]
78. Lindhagen, E.; Nygren, P.; Larsson, R. The fluorometric microculture cytotoxicity assay. *Nat. Protoc.* **2008**, *3*, 1364–1369. [CrossRef] [PubMed]
79. Pellegrin, S.; Mellor, H. The Rho family GTPase Rif induces filopodia through mDia2. *Curr. Biol.* **2005**, *15*, 129–133. [CrossRef]

 *nanomaterials*

Article

Nano-Hydroxyapatite Derived from Biogenic and Bioinspired Calcium Carbonates: Synthesis and In Vitro Bioactivity

Francesca Cestari [1,*], Francesca Agostinacchio [1,2], Anna Galotta [1], Giovanni Chemello [1], Antonella Motta [1,2,3] and Vincenzo M. Sglavo [1,3]

1   Department of Industrial Engineering, University of Trento, Via Sommarive 9, 38123 Trento, Italy; f.agostinacchio@unitn.it (F.A.); anna.galotta@unitn.it (A.G.); giovannichemello@gmail.com (G.C.); antonella.motta@unitn.it (A.M.); vincenzo.sglavo@unitn.it (V.M.S.)
2   BIOTech Research Center, and European Institute of Excellence on Tissue Engineering and Regenerative Medicine Unit, University of Trento, via delle Regole 101, 38123 Trento, Italy
3   INSTM, Via G. Giusti 9, 50121 Firenze, Italy
*   Correspondence: francesca.cestari@unitn.it

Abstract: Biogenic calcium carbonates naturally contain ions that can be beneficial for bone regeneration and therefore are attractive resources for the production of bioactive calcium phosphates. In the present work, cuttlefish bones, mussel shells, chicken eggshells and bioinspired amorphous calcium carbonate were used to synthesize hydroxyapatite nano-powders which were consolidated into cylindrical pellets by uniaxial pressing and sintering 800–1100 °C. Mineralogical, structural and chemical composition were studied by SEM, XRD, inductively coupled plasma/optical emission spectroscopy (ICP/OES). The results show that the phase composition of the sintered materials depends on the Ca/P molar ratio and on the specific $CaCO_3$ source, very likely associated with the presence of some doping elements like Mg^{2+} in eggshell and Sr^{2+} in cuttlebone. Different $CaCO_3$ sources also resulted in variable densification and sintering temperature. Preliminary in vitro tests were carried out (by the LDH assay) and they did not reveal any cytotoxic effects, while good cell adhesion and proliferation was observed at day 1, 3 and 5 after seeding through confocal microscopy. Among the different tested materials, those derived from eggshells and sintered at 900 °C promoted the best cell adhesion pattern, while those from cuttlebone and amorphous calcium carbonate showed round-shaped cells and poorer cell-to-cell interconnection.

Keywords: calcium orthophosphates; nano-hydroxyapatite; eggshell; cuttlefish bone; mussel shell; amorphous calcium carbonate

Citation: Cestari, F.; Agostinacchio, F.; Galotta, A.; Chemello, G.; Motta, A.; M. Sglavo, V. Nano-Hydroxyapatite Derived from Biogenic and Bioinspired Calcium Carbonates: Synthesis and In Vitro Bioactivity. *Nanomaterials* **2021**, *11*, 264. https://doi.org/10.3390/nano11020264

Academic Editor: Saso Ivanovski
Received: 5 January 2021
Accepted: 16 January 2021
Published: 20 January 2021

Publisher's Note: MDPI stays neutral with regard to jurisdictional claims in published maps and institutional affiliations.

1. Introduction

A material is defined as bioactive, or biologically active, when it is capable of generating an interphase bonding layer across the material–tissue interface, thus improving the ability to bond directly with the living structure. One strategy to attain bioactivity in ceramic materials is to mimic the chemistry of the bone tissue and this is why calcium phosphates and, in particular, hydroxyapatite (HA, $Ca_{10}(PO_4)_6OH_2$), have been widely studied for bone regeneration applications. Nevertheless, the main inorganic constituent of human bone is substantially different from pure HA, as it is a nanocrystalline non-stoichiometric compound containing sodium, magnesium and carbonate ions together with significant amounts of other trace elements such as K^+, F^-, Cl^-, Zn^{2+}, Sr^{2+}, Ba^{2+}, etc. For this reason, it should be more properly referred to as "impure hydroxyapatite" [1] or "biological apatite" [2]. The presence of the said additional elements has been recognized to play an important role in bone repair [3] and, for example, Mg^{2+} and Sr^{2+} have been found to improve osteoblasts adhesion and bone formation [4–8]. The impurities, especially of CO_3^{2-}, also introduce a certain degree of disorder into the crystal lattice that increases the solubility of the material and, therefore, its resorbability [9].

43

The incorporation of additional chemical elements in synthetic HA may be an expensive process, but it can be more compliant if hydroxyapatite is extracted from biological resources like mammalian bones, fish bones or biogenic calcium carbonates, which are usually in the form of calcite or aragonite [10]. Cuttlebone-derived aragonite was used to produce HA by several authors [11–17] after the pioneering work by Rocha et al. [18,19] who observed good in vitro bioactivity of HA scaffolds derived from cuttlefish bones. Other biological sources of calcium carbonate used to synthesize HA are eggshells [20–28], corals [29–31], algae [32] and sea shells [33–39]. Most of these studies report the synthesis of nanocrystalline calcium-deficient hydroxyapatite (CDHA), with significant levels of carbonate ions and other impurities, like Mg in eggshell-derived HA [40]. Another form of calcium carbonate that is found in biological systems is amorphous calcium carbonate (ACC), which can be stable or can be a transient precursor for the formation of calcite or aragonite [41], as in the case of sea urchin spines [42] and mollusk larval shells [43]. Due to its role in bio-mineralization, ACC is a promising material for bone tissue engineering, but its applicability in this field was considered in only a few studies [44,45] and, to the authors' knowledge, not for the synthesis of HA.

Although the bioactivity of naturally derived calcium phosphates has been proven by in vitro [46,47] and in vivo [48,49] biological tests, the role of calcium carbonate precursors on the biological behavior of HA is still not completely clear. Kim et al. [50] found that HA granules derived from cuttlebone promote superior cell proliferation and differentiation in vitro with respect to synthetic HA and stimulate more new bone formation in calvarial defects in white rabbits. Nevertheless, they filled the defects directly with HA granules while HA powder usually needs to be consolidated by a sintering process to develop minimal mechanical properties.

The sintering of pure HA is usually carried out at temperatures between 1100 °C and 1250 °C [51], whereas carbonate ion substitutions can lower the densification temperature to 900–950 °C [52]. Unfortunately, this thermal treatment may lead to the loss of some important properties that are thought to be beneficial for bioactivity: the carbonate content can be reduced, the nano-crystallinity lost and CDHA transformed into stoichiometric HA and β-TCP [53]. In this respect, the bioactivity of sintered synthetic HA and fish bone-derived HA was compared by Mondal et al. [54], who found that cell viability is basically the same, although osteoblasts seem to be better attached to the fish-HA than to the synthetic one.

In the present work, we synthesized HA nanopowders, starting from four different calcium carbonate sources: chicken eggshells (biogenic calcite), cuttlefish bones (biogenic aragonite), mussel shells (biogenic calcite/aragonite) and synthetic amorphous calcium carbonate (bioinspired ACC). The powders were then consolidated by uniaxial pressing and sintering at 800–1100 °C and the materials, carefully characterized, were subjected to cytotoxicity and in vitro cell adhesion tests to point out and compare their bioactivity for potential use in bone tissue engineering.

2. Materials and Methods

Chicken eggshells (ES), cuttlefish (*sepia officinalis*) bones (CB) and mussel (*mytilus galloprovincialis*) shells (MS) were collected as food waste from a local bakery, fish shop and restaurant, respectively. After washing under tap water, they were boiled in demineralized water for 10 min, dried overnight at 100 °C, ground to powder in a centrifugal mill (Retsch S100) at 400 rpm for 30 min and then sieved in order to eliminate particles larger than 300 μm. The amorphous calcium carbonate (ACC) was kindly supplied by Amorphical (Harash St. 11, Nes-Ziona, Israel) as a synthetic powder with the commercial name DENSITY™. Pure stoichiometric HA granules (sHA), powder size 5–25 μm, were achieved by S.A.I. (Science Application Industry, Saint-Priest, France).

Hydroxyapatite was synthesized from ES, CB, MS and ACC powders via wet mechanosynthesis and successive drying in an oven. The powders were mixed with ammonium phosphate dibasic ((NH_4)$_2HPO_4$, CAS: 7783-28-0, purchased from Fluka,

Buchs, Switzerland) or phosphoric acid (~85% H_3PO_4, CAS: 7664-38-2, purchased from CARLO ERBA Reagents, Cornaredo, Italy), in order to achieve a Ca/P ratio of 1.67. The mechanosynthesis reaction was promoted by ball-milling the reactants in an aqueous solution, using a Turbula® mixer and a 250 mL polyethylene bottle filled with zirconia balls (ball mass = 0.5 g), with balls-to-powder mass ratio equal to 10:1. The resulting slurry was dried in an oven for 24 h.

In a previous work the effect of the pH, the milling time and the drying temperature on the mechanosynthesis process efficiency was investigated [55]. According to the obtained results, the process parameters listed in Table 1 were selected in order to maximize the efficiency and minimize the processing time and temperature. Therefore, CB and ACC, that were more prone to be converted into HA, were ball-milled for 30 min and dried at 120 °C, while ES and MS needed 4 h milling and 150 °C drying temperature. We used $(NH_4)_2HPO_4$ as the phosphate provider for CB, MS and ACC, while ES were processed with H_3PO_4 to facilitate the reaction by the dissolution of $CaCO_3$.

Table 1. The parameters used during the mechanosynthesis process.

Raw Material	Label	Milling Time	Phosphate Provider	Drying Temperature
DENSITY™	ACC	30 min	$(NH_4)_2HPO_4$	120 °C
Eggshell	ES	4 h	H_3PO_4	150 °C
Mussel shell	MS	4 h	$(NH_4)_2HPO_4$	150 °C
Cuttlebone	CB	30 min	$(NH_4)_2HPO_4$	120 °C

The as-synthesized HA powders were then consolidated by uniaxial pressing and sintering. About 0.1 g of each powder was pressed with 2 tons in a 5 mm diameter cylindrical die using a Specac manual hydraulic press. The pellets were heated at 10 °C/min in a muffle furnace (Nabertherm P330), maintained at the selected sintering temperature for 2 h and free cooled in the oven. The sintering temperature was set for each material after some preliminary dilatometric tests on the green pellets, in order to densify the powders while retaining some porosity. The sintering temperature was set to 800 °C for ACC–HA, 900 °C for ES–HA, 1000 °C for MS–HA and 1100 °C for sHA. For CB–HA two temperatures were selected, 900 °C and 1100 °C. For practical convenience, the final materials were named after the calcium carbonate precursor and the sintering temperature as follows: ACC-800, ES-900, MS-1000, sHA-1100, CB-900 and CB-1100.

The crystalline phases of calcium precursors, as-synthesized powders and sintered pellets were characterized by x-ray diffraction (XRD). The $CaCO_3$ powders spectra were acquired in reflection geometry with an Italstructures IPD3000 X-ray diffractometer, equipped with a Co anode source (Kα radiation 1.78892 Å), a multilayer monochromator to suppress k-beta radiation, fixed 100 μm slits and an Inel CPS120 detector over 5–120° 2-theta range (0.03 degrees per channel). The as-synthesized powders and sintered pellets, instead, were analyzed using a Rigaku IIID-max, Cu anode source (Kα radiation 1.5406 Å), 5–110° 2-theta range, scan step 0.05° and step time 2 s. The spectra were then analyzed with the Rietveld software Maud [56], using the following crystal phases downloaded from the database COD [57]: calcite n. 4,502,443 [58], aragonite n. 2,100,187 [59], HA n. 4,317,043 [60], β-TCP n. 1,517,238 [61], CPP ($Ca_2P_2O_7$) n. 1,001,556 [62], CaO n. 9,006,712 [63] and CaOH n. 1,000,045 [64].

The Ca and P content and the presence of other elements in the calcium carbonate precursors and in the as-synthesized powders were determined with inductively coupled plasma/optical emission spectroscopy (ICP/OES), using Spectro Ciros Vision CCD (125–770 nm) and hydroxyapatite ultrapure standard (>99.995% trace metal basis, Sigma–Aldrich, St. Louis, MO, USA). The samples were solubilized in ultrapure nitric acid (70 vol%, Sigma–Aldrich, St. Louis, MO, USA) and diluted with pure water (obtained by reverse osmosis, $\sigma < 0.1$ μS cm^{-1}). The emission lines chosen for the analysis were 184 nm for Ca and 178 nm for P. The analyses of the other elements shall be considered as semi-quantitative, only.

The internal porosity and surface morphology of the sintered pellets were observed with a JEOL JSM-5500 scanning electron microscope (SEM) using secondary electrons, after Pt/Pd metallization with a QuorumQ150TES sputtering equipment. The density of the sintered pellets was estimated by weighting and measuring 5 samples per type, using a caliper and a laboratory scale.

The cytotoxicity of the sintered pellets was evaluated using human lung fibroblasts cell line (MRC5), expanded and cultured under standard conditions (37 °C and 5% CO_2) in minimal essential medium (MEM—Gibco, Thermo Fisher Scientific, Waltham, MA, USA), 10% of inactivated fetal bovine serum (Euroclone, Pero, Italy), supplemented with 1% of L-glutamine (Euroclone, Pero, Italy), sodium pyruvate (Gibco, Thermo Fisher Scientific, Waltham, MA, USA), non-essential amino acids (SigmaAldrich, St. Louis, MO, USA), and antibiotic-antimycotic (Euroclone, Pero, Italy). All samples were sterilized by autoclave. LDH cytotoxic assay (Thermo Fisher Scientific, Waltham, MA, USA) was performed to measure the amount of lactate dehydrogenase (LDH) released by cells during cell death. The test was performed following the European Standard EN ISO-10993-12:2004 and 10993-5:2009. Specifically, all samples without cells were incubated in medium without phenol red and with heat inactivated serum for 72 h (conditioned medium). After incubation, the conditioned medium was pursed on MRC5 cells at 70% of confluence, previously seeded at a density of 5000 cell/well in a 96-well plate and incubated for 48 h. Positive and negative controls were represented by fully lysate cells and cells cultured in standard medium, respectively. After the incubation, all samples were prepared following the manufacturer's instructions and the LDH amount released in the medium was measured using a Tecan Infinite 200 microplate reader (Tecan Group, Männedorf, Switzerland), recording the absorbance at 490 nm and the background at 680 nm. Five replicates were tested per each condition.

Cell adhesion was evaluated by confocal analysis (A1 Laser Microscope, Nikon Instruments Europe BV, Amsterdam, The Netherlands) using a human osteosarcoma cell line (MG63). The sterilized sintered pellets were placed in a 96-well culture plate and seeded with 6000 cells/well. The cells were cultured at 37 °C with 5% CO_2 in a culture media composed of MEM (Gibco, Thermo Fisher Scientific, Waltham, MA, USA), 10% fetal bovine serum (Euroclone, Pero, Italy), 1% sodium pyruvate (Gibco, Thermo Fisher Scientific, Waltham, MA, USA), 1% non-essential amino acids (Sigma-Aldrich, St. Louis, MO, USA), 1% L-glutamine (Euroclone, Pero, Italy) and 1% of antibiotic/antimycotic (Euroclone, Pero, Italy). All samples were observed at day 1, 3, and 5 after seeding. Before the confocal analysis, the specimens were fixed with 4% paraformaldehyde for 40 min at each time point, and later cell membranes were permeabilized with Triton X-100 at 0.2%. The cellular nuclei and the cytoskeleton were stained with 4, 6 diamidino 2 phenylindole, dilactate (DAPI—SigmaAldrich, St. Louis, MO, USA) and I-Fluor 488 (Abcam, Cambridge, UK), respectively.

3. Results and Discussion

3.1. Materials Characterization

The XRD spectra of the raw materials (Figure 1a) and the corresponding quantitative phase analysis (Table 2) show that eggshells (ES) are composed of 100% calcite, cuttlebones (CB) of 100% aragonite and mussel shells (MS) of a mixture of ~70/30 calcite/aragonite. The amorphous structure of ACC, instead, is revealed by the two broad bands at about 35° and 53° (Co Kα radiation) [65]. The diffraction spectra of the synthesized powders shown in Figure 1b confirm that all materials were converted into hydroxyapatite, with about 7 wt% residual calcite only in the case of MS. The width of the HA peaks indicates that the crystals are nanosized, as confirmed also by the Rietveld analyses, which pointed out crystallite sizes of 13 nm (ES–HA and MS–HA), 20 nm (ACC–HA) and 25 nm (CB–HA). In addition, the high relative intensity of the (002) peak at 25° with respect to the (211) one at 31° suggests a preferential growth along the c-axis for all materials [66].

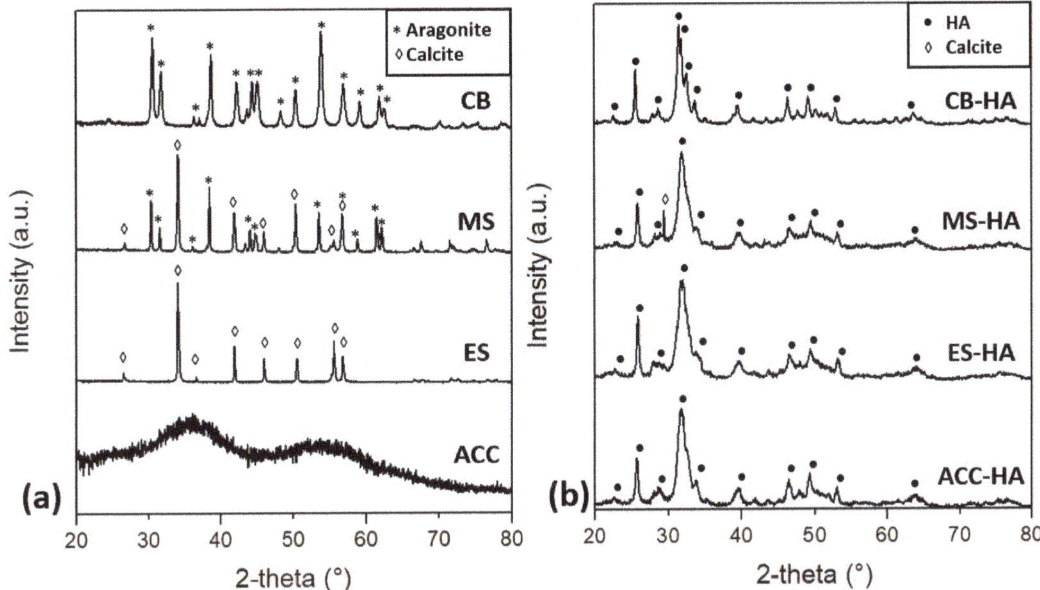

Figure 1. XRD spectra of the (**a**) raw materials and (**b**) synthesized powders.

Table 2. The phase composition (wt%) of the raw materials, synthesized powders and sintered pellets determined with Rietveld analyses.

Raw Material		Synthesized Powder		Sintered Pellet	
Label	**Phase Composition**	**Label**	**Phase Composition**	**Label**	**Phase Composition**
ACC	100% amorphous $CaCO_3$	ACC–HA	100% HA	ACC-800	~85% β-TCP, ~15% CPP
ES	100% calcite	ES–HA	100% HA	ES-900	~50% HA, ~50% β-TCP
MS	~70% calcite, ~30% aragonite	MS–HA	~93% HA, ~7% calcite	MS-1000	HA, <3% CaO
CB	100% aragonite	CB–HA	100% HA	CB-900	100% HA
				CB-1100	~90% β-TCP, ~5% HA, ~5% CaOH
sHA	-	-	-	sHA-1100	HA, <3% CaO

The sintered pellets show a different crystal structure with respect to the synthesized powders, as shown in Figure 2 and summarized in Table 2. First of all, the crystallites dimension is no longer nanometric but they become micrometric (>200 nm). MS-1000 and CB-900 maintain the hexagonal HA structure but the preferred orientation is attenuated (MS-1000) or even completely nullified (CB-900).

The results of the ICP/OES analyses are summarized in Table 3. Significant levels of Na (1.5%) and P (~2%) were found in the raw ACC powder; Mg is present (~0.4%) in the eggshell and Sr (~0.2%) in the cuttlebone. Traces of Na, Mg and Sr were detected in all biogenic $CaCO_3$, while K was found only in CB and ES. The amount of Ca and P measured in the synthesized HA powders yields to Ca/P ratio lower than 1.67 for ACC–HA, CB–HA and ES–HA, pointing out that these nanopowders are constituted by calcium deficient HA (CDHA). MS–HA and sHA, instead, are characterized by a Ca/P ratio larger than 1.67. It has to be said that, since it was not possible to completely dissolve the biogenic

HA powders in nitric acid solution during the preparation of the samples, the results may slightly deviate from the reality, and for this reason the data are presented with an estimated error of ±0.02. Other very limited impurities (<50 ppm) were also found in the materials like Cl in ACC and CB; Ba in CB and ES; Zn in CB; Fe, Si, and Mn in MS.

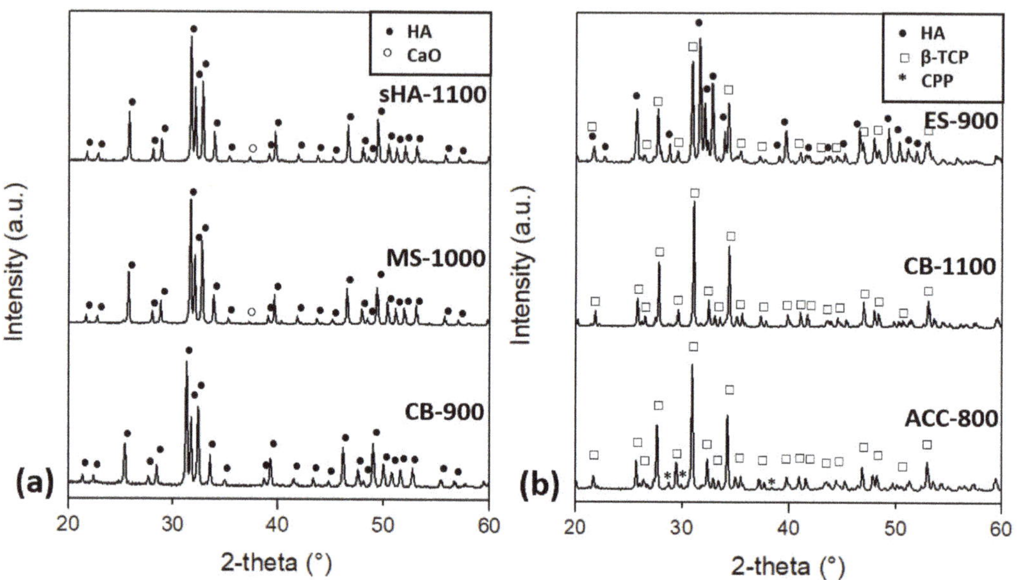

Figure 2. XRD spectra of the sintered pellets: (**a**) sHA-1100, MS-1000 and CB-900; (**b**) ES-900, CB-1100 and ACC-800.

Table 3. The concentration of the fundamental elements and Ca/P molar ratio in the raw materials ($CaCO_3$) and synthesized powders (HA) as determined by inductively coupled plasma/optical emission spectroscopy (ICP/OES) analysis.

	P	Ca/P Molar Ratio	Na		K		Mg		Sr	
	$CaCO_3$	HA	$CaCO_3$	HA	$CaCO_3$	HA	$CaCO_3$	HA	$CaCO_3$	HA
ACC	2.0%	1.28 ± 0.02	1.5%	1.4%	<0.1%	-	-	-	<0.1%	<0.1%
MS	<0.2%	1.76 ± 0.02	0.3%	0.3%	<0.1%	-	0.1%	0.1%	0.1%	0.1%
CB	<0.2%	1.64 ± 0.02	0.7%	0.9%	0.1%	0.1%	<0.1%	0.1%	0.2%	0.2%
ES	<0.2%	1.58 ± 0.02	0.1%	0.1%	0.1%	0.1%	0.4%	0.3%	<0.1%	<0.1%
sHA	-	1.71 ± 0.02	-	<0.1%	-	<0.1%	-	<0.1%	-	<0.1%

The results of the XRD analyses can be correlated with the amount of P and Ca measured by ICP/OES in the HA powders, considering the CaO/P_2O_5 phase diagram [67]. In the presence of water, 100% HA is expected with Ca/P = 1.67, while 100% β-TCP is formed with Ca/P = 1.5. For a Ca/P ratio between these two values both phases are expected following the lever rule. Accordingly, ES–HA transforms from CDHA to biphasic HA/β-TCP when sintered at 900 °C (ES-900), in the proportion expected for Ca/P = 1.58, which is 50/50. Conversely, CB–HA, with Ca/P = 1.64, maintains the hexagonal HA structure after heat treatment at 900 °C (CB-900), but transforms into ~90% β-TCP, ~5% HA and ~5% CaOH when treated at 1100 °C (CB-1100).

The different behavior of eggshell- and cuttlebone-derived HA might be related to the different content of bivalent cations, with Sr^{2+} being mainly present in CB and Mg^{2+} in ES. In fact, it is known that both magnesium and strontium stabilize the crystal structure of β-TCP, at the expense of HA [68–70]. Nevertheless, their effect is different, because cations with ionic radii larger than Ca^{2+}, such as Sr^{2+}, can be incorporated in the apatite structure to a much greater extent than those with a smaller ionic radius like Mg^{2+} [1]. Magnesium is thought to inhibit apatite crystal growth, to stabilize β-TCP with respect to α-TCP [71] and to lower the temperature at which CDHA transforms into a biphasic mixture of β-TCP and HA [72]. Therefore, the presence of Mg in ES–HA could be the reason for ES–HA to transform partially into β-TCP at 900 °C while CB–HA does not. However, it is surprising that CB–HA, when sintered at 1100 °C, is composed mainly by β-TCP, when the phase diagram predicts the prevalence of HA (~85% HA). The extra amount of calcium results instead in the presence of a small amount of CaOH. It is possible that the Sr^{2+} content caused a deviation from the expected behavior by favoring the formation of β-TCP, although the reason for the CaOH formation is still unclear. One possible explanation could be the non-homogenous distribution of Ca^{2+} and PO_4^{3-} ions, which can determine a Ca/P ratio larger than 1.67 in certain areas, leading to the formation of calcium hydroxide.

When calcium and phosphorus are present in proportions larger than 1.67, the phase diagram predicts the presence of CaO together with HA, as a result of the extra amount of calcium. Accordingly, the XRD spectra of sHA-1100 and MS-1100, whose powders are characterized by Ca/P equal to 1.71 and 1.76, respectively, revealed the presence of a small amount of CaO. Moreover, the as-synthesized MS–HA powder contains ~7% of unreacted calcite, probably because of the insufficient amount of phosphorus.

As for ACC–HA, the measured Ca/P ratio is 1.28, which falls into a region of the phase diagram where, instead of HA, a mixture of β-TCP (Ca/P = 1.5) and CPP (calcium pyrophosphate, $Ca_2P_2O_7$, Ca/P = 1.0) is expected [67]. Accordingly, ACC–HA converts into ~85 wt% β-TCP and ~15 wt% CPP upon sintering at 800 °C (ACC-800). Nevertheless, if we predict the Ca/P ratio based on the phase composition determined by the Rietveld analyses as in [73], we obtain Ca/P = 1.41 instead of 1.28. This discrepancy could be explained by considering that phosphorus is also present in the initial ACC raw powder (Table 3), most probably because it was added to stabilize the amorphous phase [74]. It is possible that said phosphorus content was detected by ICP/OES but it was not fully available for the formation of calcium phosphates, being, for example, in a form that is different from PO_4^{3-} ions. This could explain the discrepancy between the Rietveld analysis and the Ca/P ratio measured by ICP/OES, taking also into account that Ca and P in CDHA are rarely in proportions lower than 1.30 [75].

The morphology of the fracture and external surfaces of the sintered pellets are shown in Figure 3. Well-developed necks among particles are visible in all samples, although the densification is clearly not complete, as confirmed also by the density measurements reported in Table 4. As expected, CB-1100 reaches a higher densification with respect to CB-900, the relative density being ~85% and ~75%, respectively. The grains of CB-1100, visible on the material surface, look slightly bigger than those of HA-1100, even if the density is very similar. This could be correlated with the recrystallization process from CDHA to β-TCP that occurred simultaneously with the densification in CB-1100, while HA-1100 showed grain growth without recrystallization.

The density measurements and SEM analyses also show that the densification of ACC-800 is comparable to that of MS-1000 (Table 4) and its grain size is very similar to CB-900 and MS-1000 (Figure 3). The fact that ACC–HA densifies at lower temperatures with respect to biogenic HA powders could be explained by the presence of some amorphous $CaCO_3$ that could favor sintering, remaining unreacted after the synthesis process, but not being clearly revealed by XRD.

The SEM micrographs also show that, for all materials, most of the pores have dimensions of about 1 μm or less, therefore it can be assumed that they are not perceived as holes by the cells, which have dimensions close to 100 μm. However, the presence of micrometric pores is thought to be beneficial for the interaction of the bioceramics with the cells [76].

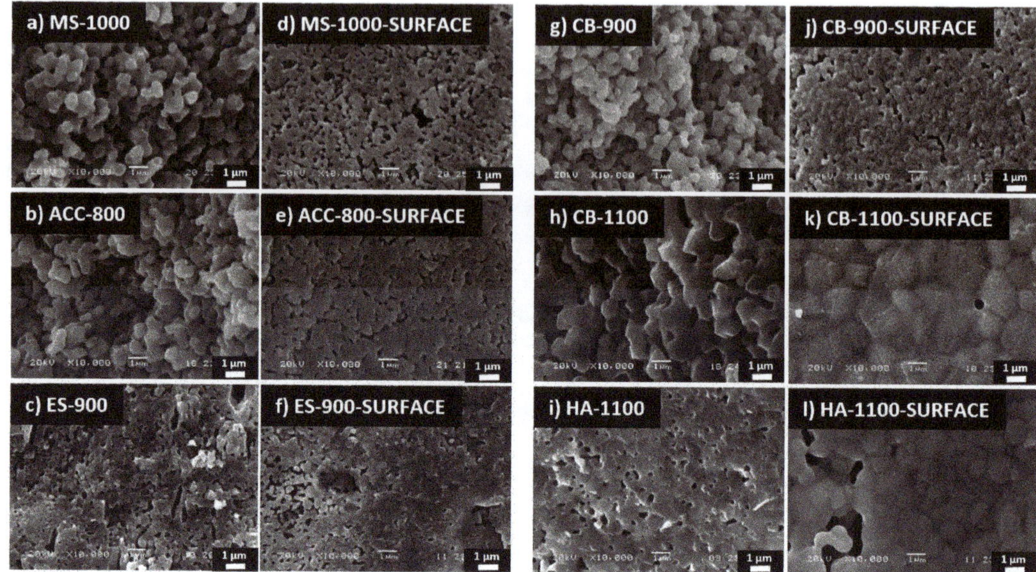

Figure 3. SEM images of the sintered materials: fracture surfaces (**a–c,g–i**); external surfaces that come into contact with the cells (**d–f,j–l**).

Table 4. The bulk and relative density of the sintered pellets.

	ACC-800	MS-1000	CB-900	CB-1100	ES-900	sHA-1100
Bulk density (g/cm^3)	2.19 ± 0.03	2.18 ± 0.08	2.33 ± 0.03	2.78 ± 0.03	2.06 ± 0.02	2.82 ± 0.08
Relative density (%)	71 ± 1	69 ± 2	74 ± 1	91 ± 1	66 ± 1	89 ± 2

3.2. In Vitro Biological Evaluation

LDH assay was performed to evaluate potential cytotoxic effects. After 48 h incubation, the conditioned medium in contact with cells was tested and the results are shown in Figure 4. According to the European Standard EN ISO-10993-12:2004 and 10993-5:2009, samples are considered cytotoxic when the LDH amount released into the medium is equal to or above 30%. As shown, all tested conditions were below the threshold of cytotoxicity. Indeed, LDH released did not exceed 8%, demonstrating that HA derived from chicken eggshells (biogenic calcite), cuttlefish bones (biogenic aragonite), mussel shells (biogenic calcite/aragonite), and synthetic amorphous calcium carbonate (bioinspired ACC) do not exhibit any cytotoxic effects on MRC5 cells.

Based on the cytotoxicity results, the same formulations were tested to study the preliminary cell adhesion at day 1, 3 and 5 after seeding, using the MG63 osteosarcoma cell line as cellular model. Cell adhesion was studied by confocal microscopy and the nuclei are stained in blue (DAPI staining) and cell cytoskeletons in green (I-Fluor488). As shown in Figure 5, at day 1, all conditions exhibited round shape cells and, among them, low cell adhesion can be detected in MS-1000 and CB-1100. At day 3, HA-1100 and ES-900 show a sensible improvement in the cell adhesion pattern, with an elongated cell morphology. Conversely, on CB-900, CB-1100, and ACC-800, at day 3, the adhesion is lower. Only the sample MS-1000 does not exhibit any differences at day 3 compared to day 1. Differently from what observed at previous times, at day 5, almost all the conditions show good cell adhesion, but with different shapes. In detail, evident cellular elongations are visible on HA-1100 sample at day 5 (clearer in the zoom, Figure 5). Cell–

cell interconnections are easily visible and some cellular clusters can also be detected. Good cell adhesion and long-shaped cytoskeletons can be observed also on CB-1100, ES-900 and MS-1000, without the formation of clusters. However, as is clear from the zoom for MS-1000 at day 5 in Figure 5, the adhered cell's morphology is clearly different from the other conditions since the cell cytoskeleton is more stretched and elongated. Lastly, CB-900 and ACC-800 are the only two samples exhibiting poor cell adhesion with the prevalence of round cellular shape.

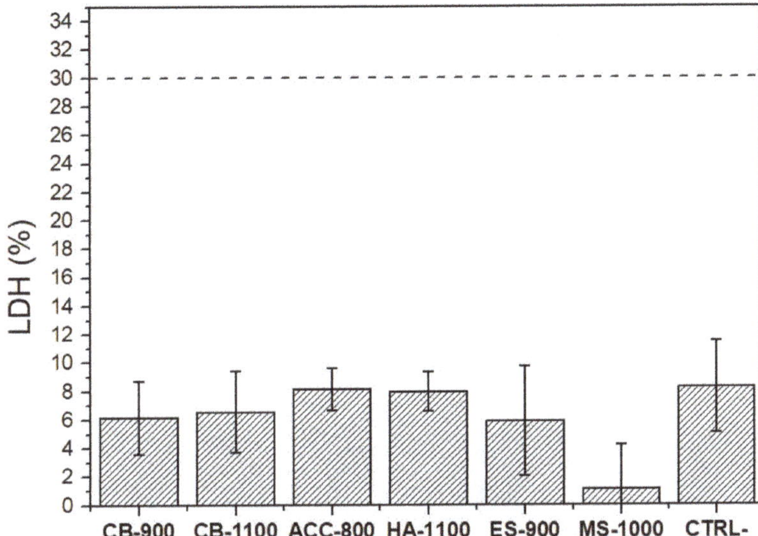

Figure 4. The LDH levels of the sintered pellets. The cytotoxicity threshold (30%) is indicated by the dashed line.

In general, in all samples cell adhesion density increases upon culture as well as cell–cell interconnections, and cells are homogeneously distributed, with the exception of HA-1100, where clusters can be observed. If the samples obtained from natural resources are compared, ES-900 induces the fastest and highest cell density and CB-900 the lowest cell–cell interconnection. Considering the cell adhesion morphology, adhered cells are less spread on ACC-800 and CB-900, with the opposite effect being shown by CB-1100, ES-900, and MS-1100.

It is well known that ions play an important role in bone biology, cell spreading and adhesiveness. Among the different ions, besides zinc and calcium, magnesium can contribute to the adhesion and spreading of MG63 cells [77,78]. ICP analyses revealed the highest concentration of magnesium on ES-900 and this might explain the good cell adhesion pattern obtained on the eggshell-derived material. Conversely, ACC-800 showed the highest concentration of phosphorous and the presence of only sodium. The absence of magnesium among the other ions might have negatively affected cell adhesion and shape. In addition, although strontium is reported to have a beneficial effect on bone structural strength and osteoblast differentiation [78], in this preliminary test only early cell adhesion was studied. For this reason, even if cuttlebone-derived samples showed the highest strontium concentration, it cannot be directly correlated with the results obtained, and differentiation studies need to be performed. Moreover, the effect of HA and β-TCP phases on cell adhesion is difficult to correlate. In detail, in spite of the fact that low cell adhesion was detected in CB-900 composed of only the HA phase, the other samples, presenting the same HA phase, showed a good cell adhesion pattern, the opposite being revealed for β-TCP. Indeed, the best cell adhesion was observed on ES-900, which is a biphasic material (50/50 HA/β-TCP). Besides ionic and phase composition, the morphology of the surfaces where cells are seeded is probably relevant. The difference between the CB-900 and CB-1100 is visible via SEM imagery, and this might be sensed by cells in a different way, affecting their behavior.

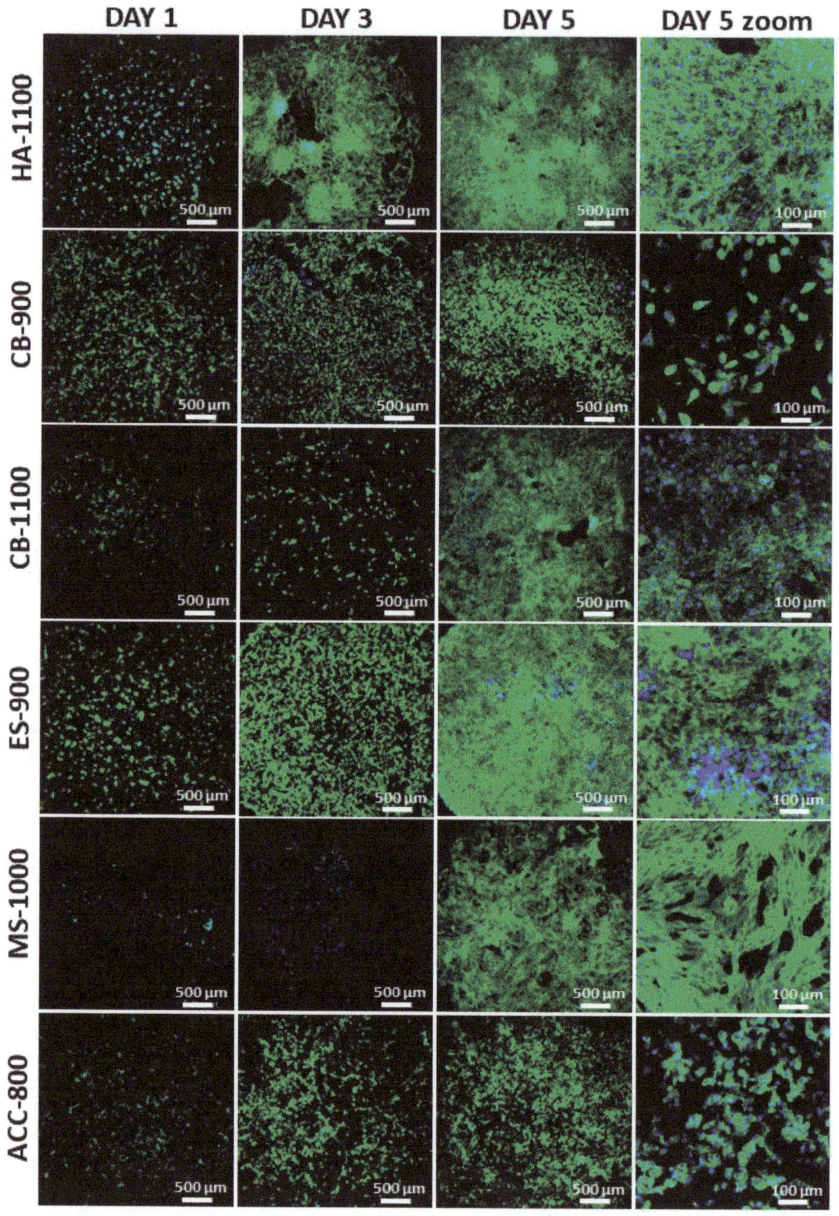

Figure 5. Confocal images of the cells adhered on the sintered pellets at day 1, 3 and 5 (the latter also with zoomed images).

4. Conclusions

Nanocrystalline hydroxyapatite derived from cuttlefish bones, eggshells, mussel shells and amorphous calcium carbonate can be synthesized and subsequently consolidated between 800 °C and 1100 °C to obtain bioactive calcium phosphate nanomaterials. The resulting crystalline phases (mainly HA, β-TCP and CPP) depend not only on the Ca/P molar ratio but also on the specific biogenic source, most probably due to the presence of different ionic species, like Mg^{2+} in eggshell-derived HA and Sr^{2+} in cuttlebone-derived HA.

The produced materials revealed the absence of any cytotoxic effect and good cell adhesion properties. Among the different calcium carbonate sources, eggshell-derived HA promotes the best cell adhesion and proliferation, which are comparable with those of pure HA, but without the formation of clusters. Cuttlebone- and mussel-derived HA also support cell adhesion well, while ACC- and cuttlebone-derived HA (sintered at 900 °C) show poor cell–cell interconnections.

In conclusion, the results show that the considered biogenically and biomimetically derived materials are valid and compatible sources for producing materials suitable for bone regeneration application.

Author Contributions: Conceptualization, F.C. and V.M.S.; methodology, F.C., V.M.S. and A.M.; formal analysis, F.C., F.A., A.G., G.C.; investigation, F.C. and F.A.; resources, V.M.S. and A.M.; data curation, F.C. and F.A.; writing—original draft preparation, F.C. and F.A.; writing—review and editing, V.M.S. and A.M.; supervision, V.M.S.; project administration, V.M.S. and A.M.; funding acquisition, V.M.S. and A.M. All authors have read and agreed to the published version of the manuscript.

Funding: This work is financially supported within the program Departments of Excellence 2018–2022 (DII-UNITN)—Project REGENERA—Italian Ministry of University and Research (MIUR).

Data Availability Statement: The data presented in this study are available on request from the corresponding author.

Acknowledgments: Amorphical (Harash St. 11, Nes-Ziona, Israel) is kindly acknowledged for providing DENSITY™ samples.

Conflicts of Interest: The authors declare no conflict of interest. The funders had no role in the design of the study; in the collection, analyses, or interpretation of data; in the writing of the manuscript, or in the decision to publish the results.

References

1. LeGeros, R.Z.; LeGeros, J.P. Phosphate Minerals in Human Tissues. *Photosphate Miner.* **1984**, 351–385. [CrossRef]
2. Dorozhkin, S.V.; Epple, M. Biological and Medical Significance of Calcium Phosphates. *Angew. Chem. Int. Ed.* **2002**, *41*, 3130–3146. [CrossRef]
3. Clarke, S.; Walsh, P. *Marine Organisms for Bone Repair and Regeneration*; Elsevier BV: Amsterdam, The Netherlands, 2014; pp. 294–318.
4. Pierantozzi, D.; Scalzone, A.; Jindal, S.; Stīpniece, L.; Šalma-Ancāne, K.; Dalgarno, K.; Gentile, P.; Mancuso, E. 3D printed Sr-containing composite scaffolds: Effect of structural design and material formulation towards new strategies for bone tissue engineering. *Compos. Sci. Technol.* **2020**, *191*, 108069. [CrossRef]
5. Liu, D.; Nie, W.; Li, D.; Wang, W.; Zheng, L.; Zhang, J.; Zhang, J.; Peng, C.; Mo, X.; He, C. 3D printed PCL/SrHA scaffold for enhanced bone regeneration. *Chem. Eng. J.* **2019**, *362*, 269–279. [CrossRef]
6. Landi, E.; Uggeri, J.; Medri, V.; Guizzardi, S. Sr, Mg cosubstituted HA porous macro-granules: Potentialities as resorbable bone filler with antiosteoporotic functions. *J. Biomed. Mater. Res. Part A* **2013**, *101*, 2481–2490. [CrossRef]
7. Sader, M.S.; LeGeros, R.Z.; Soares, G.A. Human osteoblasts adhesion and proliferation on magnesium-substituted tricalcium phosphate dense tablets. *J. Mater. Sci. Mater. Med.* **2009**, *20*, 521–527. [CrossRef]
8. Holzapfel, B.M.; Reichert, J.C.; Schantz, J.-T.; Gbureck, U.; Rackwitz, L.; Nöth, U.; Jakob, F.; Rudert, M.; Groll, J.; Hutmacher, D.W. How smart do biomaterials need to be? A translational science and clinical point of view. *Adv. Drug Deliv. Rev.* **2013**, *65*, 581–603. [CrossRef]
9. Supová, M. Isolation and preparation of nanoscale bioapatites from natural sources: A review. *J. Nanosci. Nanotechnol.* **2014**, *14*, 546–563. [CrossRef]
10. Akram, M.; Ahmed, R.; Shakir, I.; Ibrahim, W.A.W.; Hussain, R. Extracting hydroxyapatite and its precursors from natural resources. *J. Mater. Sci.* **2014**, *49*, 1461–1475. [CrossRef]

11. Lee, S.J.; Lee, Y.C.; Yoon, Y.S. Characteristics of calcium phosphate powders synthesized from cuttlefish bone and phosphoric acid. *J. Ceram. Process. Res.* **2007**, *8*, 427–430.

12. Ivankovic, H.; Tkalčec, E.; Orlić, S.; Ferrer, G.G.; Schauperl, Z. Hydroxyapatite formation from cuttlefish bones: Kinetics. *J. Mater. Sci. Mater. Electron.* **2010**, *21*, 2711–2722. [CrossRef]

13. Ivankovic, H.; Ferrer, G.G.; Tkalcec, E.; Orlic, S. Preparation of highly porous hydroxyapatite from cuttlefish bone. *J. Mater. Sci. Mater. Med.* **2009**, *20*, 1039–1046. [CrossRef]

14. Tkalčec, E.; Popović, J.; Orlić, S.; Milardović, S.; Ivanković, H. Hydrothermal synthesis and thermal evolution of carbonate-fluorhydroxyapatite scaffold from cuttlefish bones. *Mater. Sci. Eng. C* **2014**, *42*, 578–586. [CrossRef]

15. Reinares-Fisac, D.; Veintemillas-Verdaguer, S.; Fernández-Díaz, L. Conversion of biogenic aragonite into hydroxyapatite scaffolds in boiling solutions. *CrystEngComm* **2016**, *19*, 110–116. [CrossRef]

16. Venkatesan, J.; Rekha, P.D.; Anil, S.; Bhatnagar, I.; Sudha, P.N.; Dechsakulwatana, C.; Kim, S.-K.; Shim, M.S. Hydroxyapatite from Cuttlefish Bone: Isolation, Characterizations, and Applications. *Biotechnol. Bioprocess Eng.* **2018**, *23*, 383–393. [CrossRef]

17. Milovac, D.; Ferrer, G.G.; Ivankovic, M.; Ivankovic, H. PCL-coated hydroxyapatite scaffold derived from cuttlefish bone: Morphology, mechanical properties and bioactivity. *Mater. Sci. Eng. C* **2014**, *34*, 437–445. [CrossRef]

18. Rocha, J.; Lemos, A.; Agathopoulos, S.; Valério, P.; Kannan, S.; Oktar, F.; Ferreira, J.M. Scaffolds for bone restoration from cuttlefish. *Bone* **2005**, *37*, 850–857. [CrossRef]

19. Kannan, S.; Rocha, J.; Agathopoulos, S.; Ferreira, J. Fluorine-substituted hydroxyapatite scaffolds hydrothermally grown from aragonitic cuttlefish bones. *Acta Biomater.* **2007**, *3*, 243–249. [CrossRef]

20. Rivera, E.M.; Araiza, M.; Brostow, W.; Castaño, V.M.; Díaz-Estrada, J.; Hernández, R.; Rodríguez, J. Synthesis of hydroxyapatite from eggshells. *Mater. Lett.* **1999**, *41*, 128–134. [CrossRef]

21. Lee, S.; Oh, S. Fabrication of calcium phosphate bioceramics by using eggshell and phosphoric acid. *Mater. Lett.* **2003**, *57*, 4570–4574. [CrossRef]

22. Balázsi, C.; Wéber, F.; Kövér, Z.; Horváth, E.; Németh, C. Preparation of calcium–phosphate bioceramics from natural resources. *J. Eur. Ceram. Soc.* **2007**, *27*, 1601–1606. [CrossRef]

23. Sanosh, K.; Chu, M.-C.; Balakrishnan, A.; Kim, T.-N.; Cho, S.-J. Utilization of biowaste eggshells to synthesize nanocrystalline hydroxyapatite powders. *Mater. Lett.* **2009**, *63*, 2100–2102. [CrossRef]

24. Gergely, G.; Wéber, F.; Lukács, I.; Tóth, A.L.; Horváth, Z.E.; Mihály, J.; Balázsi, C. Preparation and characterization of hydroxyapatite from eggshell. *Ceram. Int.* **2010**, *36*, 803–806. [CrossRef]

25. Goloshchapov, D.; Kashkarov, V.; Rumyantseva, N.; Domashevskaya, E.P.; Lenshin, A.; Agapov, B.; Domashevskaya, E. Synthesis of nanocrystalline hydroxyapatite by precipitation using hen's eggshell. *Ceram. Int.* **2013**, *39*, 4539–4549. [CrossRef]

26. Ho, W.-F.; Hsu, H.-C.; Hsu, S.-K.; Hung, C.-W.; Wu, S.-C. Calcium phosphate bioceramics synthesized from eggshell powders through a solid state reaction. *Ceram. Int.* **2013**, *39*, 6467–6473. [CrossRef]

27. Ramesh, S.; Natasha, A.; Tan, C.; Bang, L.; Ching, C.; Chandran, H. Direct conversion of eggshell to hydroxyapatite ceramic by a sintering method. *Ceram. Int.* **2016**, *42*, 7824–7829. [CrossRef]

28. Krishna, D.S.R.; Siddharthan, A.; Seshadri, S.K.; Kumar, T.S. A novel route for synthesis of nanocrystalline hydroxyapatite from eggshell waste. *J. Mater. Sci. Mater. Med.* **2007**, *18*, 1735–1743. [CrossRef]

29. Sivakumar, M.; Kumar, T.; Shantha, K.; Rao, K. Development of hydroxyapatite derived from Indian coral. *Biomaterials* **1996**, *17*, 1709–1714. [CrossRef]

30. Jinawath, S.; Polchai, D.; Yoshimura, M. Low-temperature, hydrothermal transformation of aragonite to hydroxyapatite. *Mater. Sci. Eng. C* **2002**, *22*, 35–39. [CrossRef]

31. Hu, J.; Russell, J.J.; Bennissan, B.; Vago, R. Production and analysis of hydroxyapatite from Australian corals via hydrothermal process. *J. Mater. Sci. Lett.* **2001**, *20*, 85–87. [CrossRef]

32. Walsh, P.; Buchanan, F.; Dring, M.; Maggs, C.; Bell, S.; Walker, G. Low-pressure synthesis and characterisation of hydroxyapatite derived from mineralise red algae. *Chem. Eng. J.* **2008**, *137*, 173–179. [CrossRef]

33. Lemos, A.; Rocha, J.; Quaresma, S.; Kannan, S.; Oktar, F.; Agathopoulos, S.; Ferreira, J. Hydroxyapatite nano-powders produced hydrothermally from nacreous material. *J. Eur. Ceram. Soc.* **2006**, *26*, 3639–3646. [CrossRef]

34. De Paula, S.M.; Huila, M.; Araki, K.; Toma, H.E. Confocal Raman and electronic microscopy studies on the topotactic conversion of calcium carbonate from Pomacea lineate shells into hydroxyapatite bioceramic materials in phosphate media. *Micron* **2010**, *41*, 983–989. [CrossRef]

35. Wu, S.-C.; Hsu, H.-C.; Wu, Y.-N.; Ho, W.-F. Hydroxyapatite synthesized from oyster shell powders by ball milling and heat treatment. *Mater. Charact.* **2011**, *62*, 1180–1187. [CrossRef]

36. Rujitanapanich, S.; Kumpapan, P.; Wanjanoi, P. Synthesis of Hydroxyapatite from Oyster Shell via Precipitation. *Energy Procedia* **2014**, *56*, 112–117. [CrossRef]

37. Shavandi, A.; Bekhit, A.E.-D.A.; Ali, A.; Sun, Z. Synthesis of nano-hydroxyapatite (nHA) from waste mussel shells using a rapid microwave method. *Mater. Chem. Phys.* **2015**, *149–150*, 607–616. [CrossRef]

38. Pal, A.; Maity, S.; Chabri, S.; Bera, S.; Chowdhury, A.R.; Das, M.; Sinha, A. Mechanochemical synthesis of nanocrystalline hydroxyapatite from Mercenaria clam shells and phosphoric acid. *Biomed. Phys. Eng. Express* **2017**, *3*, 015010. [CrossRef]

39. Kumar, G.S.; Girija, E.K.; Venkatesh, M.; Karunakaran, G.; Kolesnikov, E.; Kuznetsov, D. One step method to synthesize flower-like hydroxyapatite architecture using mussel shell bio-waste as a calcium source. *Ceram. Int.* **2017**, *43*, 3457–3461. [CrossRef]

40. Kumar, G.S.; Thamizhavel, A.; Girija, E.K. Microwave conversion of eggshells into flower-like hydroxyapatite nanostructure for biomedical applications. *Mater. Lett.* **2012**, *76*, 198–200. [CrossRef]

41. Addadi, L.; Raz, S.; Weiner, S. Taking Advantage of Disorder: Amorphous Calcium Carbonate and Its Roles in Biomineralization. *Adv. Mater.* **2003**, *15*, 959–970. [CrossRef]

42. Albéric, M.; Stifler, C.A.; Zou, Z.; Sun, C.-Y.; Killian, C.E.; Valencia, S.; Mawass, M.-A.; Bertinetti, L.; Gilbert, P.U.; Politi, Y. Growth and regrowth of adult sea urchin spines involve hydrated and anhydrous amorphous calcium carbonate precursors. *J. Struct. Biol. X* **2019**, *1*, 100004. [CrossRef]

43. Weiss, I.M.; Tuross, N.; Addadi, L.; Weiner, S. Mollusc larval shell formation: Amorphous calcium carbonate is a precursor phase for aragonite. *J. Exp. Zoo L.* **2002**, *293*, 478–491. [CrossRef]

44. Jia, S.; Guo, Y.; Zai, W.; Su, Y.; Yuan, S.; Yu, X.; Li, G. Preparation and characterization of a composite coating composed of polycaprolactone (PCL) and amorphous calcium carbonate (ACC) particles for enhancing corrosion resistance of magnesium implants. *Prog. Org. Coat.* **2019**, *136*, 105225. [CrossRef]

45. Myszka, B.; Schüßler, M.; Hurle, K.; Demmert, B.; Detsch, R.; Boccaccini, A.R.; Wolf, S.E. Phase-specific bioactivity and altered Ostwald ripening pathways of calcium carbonate polymorphs in simulated body fluid. *RSC Adv.* **2019**, *9*, 18232–18244. [CrossRef]

46. Cozza, N.; Monte, F.; Bonani, W.; Aswath, P.; Motta, A.; Migliaresi, C. Bioactivity and mineralization of natural hydroxyapatite from cuttlefish bone and Bioglass ® co-sintered bioceramics. *J. Tissue Eng. Regen. Med.* **2018**, *12*, e1131–e1142. [CrossRef]

47. Milovac, D.; Gamboa-Martínez, T.C.; Ivankovic, M.; Ferrer, G.G.; Ivankovic, H. PCL-coated hydroxyapatite scaffold derived from cuttlefish bone: In vitro cell culture studies. *Mater. Sci. Eng. C* **2014**, *42*, 264–272. [CrossRef]

48. Vecchio, K.S.; Zhang, X.; Massie, J.B.; Wang, M.; Kim, C.W. Conversion of sea urchin spines to Mg-substituted tricalcium phosphate for bone implants. *Acta Biomater.* **2007**, *3*, 785–793. [CrossRef]

49. Kim, B.-S.; Yang, S.-S.; Lee, J. A polycaprolactone/cuttlefish bone-derived hydroxyapatite composite porous scaffold for bone tissue engineering. *J. Biomed. Mater. Res. Part B Appl. Biomater.* **2013**, *102*, 943–951. [CrossRef]

50. Kim, B.-S.; Kang, H.J.; Yang, S.-S.; Lee, J. Comparison of in vitro and in vivo bioactivity: Cuttlefish-bone-derived hydroxyapatite and synthetic hydroxyapatite granules as a bone graft substitute. *Biomed. Mater.* **2014**, *9*, 025004. [CrossRef]

51. Champion, E. Sintering of calcium phosphate bioceramics. *Acta Biomater.* **2013**, *9*, 5855–5875. [CrossRef]

52. Landi, E.; Celotti, G.; Logroscino, G.; Tampieri, A. Carbonated hydroxyapatite as bone substitute. *J. Eur. Ceram. Soc.* **2003**, *23*, 2931–2937. [CrossRef]

53. Dorozhkina, E.I.; Dorozhkin, S.V. Mechanism of the Solid-State Transformation of a Calcium-Deficient Hydroxyapatite (CDHA) into Biphasic Calcium Phosphate (BCP) at Elevated Temperatures. *Chem. Mater.* **2002**, *14*, 4267–4272. [CrossRef]

54. Mondal, S.; Hoang, G.; Manivasagan, P.; Moorthy, M.S.; Kim, H.H.; Phan, T.T.V.; Oh, J. Comparative characterization of biogenic and chemical synthesized hydroxyapatite biomaterials for potential biomedical application. *Mater. Chem. Phys.* **2019**, *228*, 344–356. [CrossRef]

55. Cestari, F.; Chemello, G.; Galotta, A.; Sglavo, V.M. Low-temperature synthesis of nanometric apatite from biogenic sources. *Ceram. Int.* **2020**. [CrossRef]

56. Bortolotti, M.; Lutterotti, L.; Pepponi, G. Combining XRD and XRF analysis in one Rietveld-like fitting. *Powder Diffr.* **2017**, *32*, S225–S230. [CrossRef]

57. Gražulis, S.; Chateigner, D.; Downs, R.T.; Yokochi, A.F.T.; Quirós, M.; Lutterotti, L.; Manakova, E.; Butkus, J.; Moeck, P.; Le Bail, A. Crystallography Open Database—An open-access collection of crystal structures. *J. Appl. Cryst.* **2009**, *42*, 726–729. [CrossRef]

58. Zolotoyabko, E.; Caspi, E.N.; Fieramosca, J.S.; Von Dreele, R.B.; Marin, F.; Mor, G.; Addadi, L.; Weiner, S.; Politi, Y. Differences between Bond Lengths in Biogenic and Geological Calcite. *Cryst. Growth Des.* **2010**, *10*, 1207–1214. [CrossRef]

59. Caspi, E.N.; Pokroy, B.; Lee, P.L.; Quintana, J.P.; Zolotoyabko, E. On the structure of aragonite. *Acta Cryst. Sect. B Struct. Sci.* **2005**, *61*, 129–132. [CrossRef]

60. Ardanova, L.I.; Get'Man, E.I.; Loboda, S.N.; Prisedskii, V.V.; Tkachenko, T.V.; Marchenko, V.I.; Antonovich, V.P.; Chivireva, N.A.; Chebishev, K.A.; Lyashenko, A.S. Isomorphous Substitutions of Rare Earth Elements for Calcium in Synthetic Hydroxyapatites. *Inorg. Chem.* **2010**, *49*, 10687–10693. [CrossRef] [PubMed]

61. Yashima, M.; Sakai, A.; Kamiyama, T.; Hoshikawa, A. Crystal structure analysis of β-tricalcium phosphate $Ca_3(PO_4)_2$ by neutron powder diffraction. *J. Solid State Chem.* **2003**, *175*, 272–277. [CrossRef]

62. Boudin, S.; Grandin, A.; Borel, M.M.; LeClaire, A.; Raveau, B. Redetermination of the β-$Ca_2P_2O_7$ structure. *Acta Cryst. Sect. C Cryst. Struct. Commun.* **1993**, *49*, 2062–2064. [CrossRef]

63. Fiquet, G.; Richet, P.; Montagnac, G. High-temperature thermal expansion of lime, periclase, corundum and spinel. *Phys. Chem. Min.* **1999**, *27*, 103–111. [CrossRef]

64. Busing, W.R.; Levy, H.A. Neutron Diffraction Study of Calcium Hydroxide. *J. Chem. Phys.* **1957**, *26*, 563–568. [CrossRef]

65. Rodriguez-Blanco, J.D.; Shaw, S.; Benning, L.G. The kinetics and mechanisms of amorphous calcium carbonate (ACC) crystallization to calcite, viavaterite. *Nanoscale* **2010**, *3*, 265–271. [CrossRef] [PubMed]

66. Müller, L.; Conforto, E.; Caillard, D.; Müller, F.A. Biomimetic apatite coatings—Carbonate substitution and preferred growth orientation. *Biomol. Eng.* **2007**, *24*, 462–466. [CrossRef] [PubMed]

67. Gross, K.A.; Berndt, C.C. Thermal processing of hydroxyapatite for coating production. *J. Biomed. Mater. Res.* **1998**, *39*, 580–587. [CrossRef]

68. Bigi, A.; Marchetti, F.; Ripamonti, A.; Roveri, N.; Foresti, E. Magnesium and strontium interaction with carbonate-containing hydroxyapatite in aqueous medium. *J. Inorg. Biochem.* **1981**, *15*, 317–327. [CrossRef]

69. Bigi, A. Isomorphous substitutions in β-tricalcium phosphate: The different effects of zinc and strontium. *J. Inorg. Biochem.* **1997**, *66*, 259–265. [CrossRef]

70. Kannan, S.; Pina, S.; Ferreira, J.M. Formation of Strontium-Stabilized?-Tricalcium Phosphate from Calcium-Deficient Apatite *J. Am. Ceram. Soc.* **2006**, *89*, 3277–3280. [CrossRef]

71. Frasnelli, M.; Sglavo, V.M. Effect of Mg2+ doping on beta–alpha phase transition in tricalcium phosphate (TCP) bioceramics *Acta Biomater.* **2016**, *33*, 283–289. [CrossRef]

72. Batra, U.; Kapoor, S.; Sharma, S. Influence of Magnesium Ion Substitution on Structural and Thermal Behavior of Nanodimensional Hydroxyapatite. *J. Mater. Eng. Perform.* **2013**, *22*, 1798–1806. [CrossRef]

73. Ishikawa, K.; Ducheyne, P.; Radin, S. Determination of the Ca/P ratio in calcium-deficient hydroxyapatite using X-ray diffraction analysis. *J. Mater. Sci. Mater. Electron.* **1993**, *4*, 165–168. [CrossRef]

74. Aizenberg, J.; Weiner, S.; Addadi, L. Coexistence of Amorphous and Crystalline Calcium Carbonate in Skeletal Tissues. *Connect. Tissue Res.* **2003**, *44*, 20–25. [CrossRef] [PubMed]

75. Lilley, K.J.; Gbureck, U.; Wright, A.J.; Farrar, D.F.; Barralet, J.E. Cement from nanocrystalline hydroxyapatite: Effect of calcium phosphate ratio. *J. Mater. Sci. Mater. Med.* **2005**, *16*, 1185–1190. [CrossRef]

76. Turnbull, G.; Clarke, J.; Picard, F.; Riches, P.; Jia, L.; Han, F.; Li, B.; Shu, W. 3D bioactive composite scaffolds for bone tissue engineering. *Bioact. Mater.* **2018**, *3*, 278–314. [CrossRef]

77. Paul, W.; Sharma, C.P. Effect of calcium, zinc and magnesium on the attachment and spreading of osteoblast like cells onto ceramic matrices. *J. Mater. Sci. Mater. Electron.* **2006**, *18*, 699–703. [CrossRef]

78. Scalera, F.; Palazzo, B.; Barca, A.; Gervaso, F. Sintering of magnesium-strontium doped hydroxyapatite nanocrystals: Towards the production of 3D biomimetic bone scaffolds. *J. Biomed. Mater. Res. Part A* **2019**, *108*, 633–644. [CrossRef]

 *nanomaterials*

Article

Fabrication and Characterization of Biodegradable Gelatin Methacrylate/Biphasic Calcium Phosphate Composite Hydrogel for Bone Tissue Engineering

Ji-Bong Choi [1,†], Yu-Kyoung Kim [1,†], Seon-Mi Byeon [2], Jung-Eun Park [1], Tae-Sung Bae [1], Yong-Seok Jang [1,*] and Min-Ho Lee [1,*]

[1]   Department of Dental Biomaterials, Institute of Biodegradable Materials, School of Dentistry, Jeonbuk National University, Jeonju-si 54896, Jeollabuk-do, Korea; submissi@naver.com (J.-B.C.); yk0830@naver.com (Y.-K.K.); pje312@naver.com (J.-E.P.); bts@jbnu.ac.kr (T.-S.B.)
[2]   Dental Clinic of Ebarun, Suncheon-si 57999, Jeollanam-do, Korea; sumse1205@naver.com
*   Correspondence: Yjang@jbnu.ac.kr (Y.-S.J.); mh@jbnu.ac.kr (M.-H.L.)
†   These authors contributed equally to the paper.

Abstract: In the field of bone tissue, maintaining adequate mechanical strength and tissue volume is an important part. Recently, biphasic calcium phosphate (BCP) was fabricated to solve the shortcomings of hydroxyapatite (HA) and beta-tricalcium phosphate (β-TCP), and it is widely studied in the field of bone-tissue engineering. In this study, a composite hydrogel was fabricated by applying BCP to gelatin methacrylate (GelMA). It was tested by using a mechanical tester, to characterize the mechanical properties of the prepared composite hydrogel. The fabricated BCP was analyzed through FTIR and XRD. As a result, a different characteristic pattern from hydroxyapatite (HA) and beta-tricalcium phosphate (β-TCP) was observed, and it was confirmed that it was successfully bound to the hydrogel. Then, the proliferation and differentiation of preosteoblasts were checked to evaluate cell viability. The analysis results showed high cell viability and relatively high bone differentiation ability in the composite hydrogel to which BCP was applied. These features have been shown to be beneficial for bone regeneration by maintaining the volume and shape of the hydrogel. In addition, hydrogels can be advantageous for clinical use, as they can shape the structure of the material for custom applications.

Keywords: hydrogel; tissue engineering; biphasic calcium phosphate nanoparticle (BCP-NP); biocompatibility; biodegradable; gelatin methacryloyl (GelMA); visible light

Citation: Choi, J.-B.; Kim, Y.-K.; Byeon, S.-M.; Park, J.-E.; Bae, T.-S.; Jang, Y.-S.; Lee, M.-H. Fabrication and Characterization of Biodegradable Gelatin Methacrylate/Biphasic Calcium Phosphate Composite Hydrogel for Bone Tissue Engineering. *Nanomaterials* **2021**, *11*, 617. https://doi.org/10.3390/nano11030617

Academic Editor: Jean-Marie Nedelec

Received: 21 January 2021
Accepted: 15 February 2021
Published: 2 March 2021

Publisher's Note: MDPI stays neutral with regard to jurisdictional claims in published maps and institutional affiliations.

1. Introduction

Currently, the most common method for repairing damaged bone tissue is to implant bone-graft material directly into the defect. Additionally, in order to recover the defective area in the clinical environment, bone tissue is regenerated, using guided bone regeneration (GBR), and then treatment through implants is performed. Clinically available bone-graft materials are often provided in powder form. These powdered implants are mixed with saline or blood, to maintain their shape. The powdered bone-graft material can be easily applied to small bone defects. However, In the case of large defects, it is difficult to maintain the shape, due to weak adhesion by blood or saline solution. In these situations, auxiliary materials such as metal (titanium, magnesium, etc.) mesh are used to maintain shape. However, there are still some limitations related to the prolongation of the infection and recovery period after treatment and the delivery of biological mediators for tissue regeneration [1,2]. Membranes based on natural hydrogels show great potential in tissue-engineering applications, due to their tunable physical properties. In addition, the natural hydrogel-based membrane is in the spotlight, as an alternative for the delivery of various controlled substances in vivo due to micro-level control, increase in mechanical properties,

57

and control over decomposition rate. For example, a study [3] on dental regeneration by encapsulating cells fabricated by using a dentin-derived hydrogel was reported. Another study [4] reported the development of hydrogels based on photopolymerized gelatin containing cells by visible light (VL), using a dental transition device. Gelatin methacryloyl (GelMA), manufactured by substituting natural polymers, is a compound synthesized by adding a methacrylate group to the amine and hydroxyl groups of gelatins. However, in addition to the biocompatibility and expansion ratio, which are basic characteristics of gelatin, it has a motif that is degraded by RGD peptide (arginine-glycine-aspartic acid) and Matrix Metalloproteinase (MMP) related to cell adhesion [5,6]. Gelatin-methacrylic anhydride (GelMA), in which a methacrylate group is added to the gelatin amine group, can be used to manufacture a hydrogel by photocrosslinking. GelMA hydrogel using photo-crosslinking showed good biocompatibility in the body, due to its high biological properties [7]. In addition, it is widely used in tissue engineering, through cell culture and drug delivery, in consideration of properties similar to the extracellular matrix (ECM) [8,9]. However, compared to other polymers, such as alginate and polylactic acid hydrogel, the mechanical properties are significantly lower [10–12]. Therefore, in order to apply GelMA to bone-tissue regeneration, it is necessary to improve the mechanical strength by adding other materials. GelMA nanocomposites were developed by using synthetic bones, such as hydroxyapatite (HA) or β-tricalcium phosphate (β-TCP), and many studies have been conducted to improve bone function [13,14]. HA is known as the main component of bones and has high biocompatibility and strength, which complements the strength of other minerals (β-TCP). The composite material containing HA particles and HA showed good regeneration ability in the bone-defect model, but after degradation, HA remained in the regenerated site and could not be completely replaced by new bone [15]. Moreover, delayed biodegradation can interfere with bone remodeling [16]. On the contrary, β-TCP can be easily degraded in the body, activates bone conduction through micropores inside the body to form new bone, and is known to cause complete bone regeneration, because it does not remain in the tissue [17]. Biphasic calcium phosphate nanoparticles (BCP-NPs), which are a mixture of HA and β-TCP at 60/40 wt%, are known as a composite material that can compensate for the disadvantages of HA and β-TCP, which are synthetic bone materials [18]. Recently, a hydrogel system containing biphasic calcium phosphate has been studied for effective bone regeneration. Lei Nie et al. reported that BCP-NP-conjugated chitosan/gelatin hydrogel, combined with the bone marrow mesenchymal stem, exhibits an excellent bone formation reaction. Omar Faruq et al. demonstrated excellent osseointegration and improved bone formation by incorporating BCP-NP into hyaluronic acid/gelatin hydrogels [19,20]. However, further studies to confirm the effect of these hydrogel systems directly are required. In this study, we designed composite hydrogel after directly fabricating BCP-NP and combining it with the GelMA hydrogel. The composite hydrogel was chemically crosslinked by photopolymerization, using visible light, and formed a physical bond with BCP-NP. The prepared composite hydrogel was analyzed for physical and chemical properties, and biocompatibility was confirmed through an in vitro test.

2. Materials and Methods

2.1. Materials

All materials used in the synthesis of GelMA (Type A gelatin, Methacrylic anhydride (MA), dialysis tubing, high-retention seamless cellulose tubing (12–14 kDA MWCO, 40 mm diameter), Dulbecco's phosphate-buffered saline (DPBS) Dialysis tubing closures), fabrication of hydrogel (Triethanolamine (TEA), Eosin Y disodium salt, N-Vinylcaprolactam (VC)), fabrication of biphasic calcium phosphate nanoparticles (BCP-NP) (calcium nitrate tetrahydrate ($Ca(NO_3)_2 \cdot 4H_2O$, and ammonium phosphate dibasic ((NH_4)2HPO_4) were purchased from Sigma-Aldrich (St. Louis, MO, USA) and used without further purification. Moreover, α-MEM (Minimum Essential Medium Eagle-α), 10% fetal bovine serum (FBS), and antibiotics (penicillin/streptomycin) used for the cell experiment were purchased from Gibco BRL (Invitrogen Co., Carlsbad, CA, USA). Cell proliferation and activity were evaluated, using CCKi-8 (Enzo Life Science Inc., New York, NY, USA) and a TRACP & ALP assay kit (TakaRa, Kyoto City, Kyoto Prefecture, Japan).

2.2. Equipment

To confirm the substitution of GelMA, a nuclear magnetic resonance (1H-NMR) spectrum was recorded, using JNM-ECZ600R (JEOL, Akishima City, Tokyo, Japan) installed in the Center for University-Wide Research Facilities (CURF) at Jeonbuk National University. Fourier transform infrared spectroscopy (FTIR, Frontier (Perkin Elmer, Waltham, MA , USA)) was used to measure the molecular bonding structures and spectroscopic properties of hydrogels and other materials. The crystal phase of BCP-NP and polydopamine contained in the hydrogel was analyzed qualitatively and quantitatively. This was graphically plotted by using X'PERT-PRO Powder (PANalytical, Malvern, British). The crystal phase of BCP-NP was observed, using a Bio Transmission Electron Microscope (H-7650, HITACHI, Marunouchi, Chiyoda-ku, Tokyo, Japan), at a magnification of 100,000–200,000, with 100 kV of acceleration voltage. The hydrogel morphology was analyzed, using a Field-Emission Scanning Electron Microscope (FE-SEM; SU-70, HITACHI, Japan) and energy-dispersive X-ray (EDX) spectrometer (OCTANE PRO, AMETEK®, Burwin, PA, USA). The mechanical properties of the hydrogel were evaluated by using a universal testing machine (Instron 5569, Instron, Norwood, MA, USA), and cell proliferation and activity were evaluated, using an ELISA reader (Molecular Devices, EMax, San Jose, CA, USA). Additionally, thermogravimetric analysis was evaluated by using SDT Q600 (WATERS, Milford, MA, USA).

2.3. Synthesis of GelMA and BCP-NP

GelMA was synthesized by previously reported methods [8,21], using gelatin type A. Briefly, 10% (w/v) of gelatin was added to DPBS and dissolved completely at 60 °C. Then, MA (0.8 mL MA to 1 g gelatin ratio) was added to the gelatin solution and stirred at 50 °C for 3 h. The reaction was terminated by diluting the solution 5 times by volume. This was followed by dialyzing for 1 week, using a 12–14 kDA cutoff dialysis pack in deionized water, at 40 °C. The dialyzed GelMA solution was freeze-dried, at −78 °C, for 5 days, and the resulting white GelMA foam was stored in a freezer, at −20 °C, before use. The preparation method for BCP-NP is as follows [22]. First, 75.0 mL of 0.5 M calcium nitrate tetrahydrate (Ca (NO$_3$)$_2$·4H$_2$O, Sigma-Aldrich (St. Louis, MO, USA)) solution was slowly added to 50 mL of 0.5 M ammonium phosphate dibasic ((NH$_4$)$_2$HPO$_4$, Sigma-Aldrich (St. Louis, MO, USA)). The mixture was adjusted to a 9.5 pH and allowed to react at 55 °C for 30 min. The resulting slurry was aged for 40 h at 25 °C, to form stable BCP-NP. Then, BCP-NPs were centrifuged, freeze-dried, and stored in a freezer, at −20 °C, before use.

2.4. Preparation of GelMA/BCP-NP Composite Hydrogel

GelMA hydrogel was fabricated by visible (VL), using photopolymerization with Optilux Demetron curing light (Kerr Inc., Danbury, CT, USA). In distilled water containing TEA 1.88% (v/v) and VC 1.25% (w/v), different concentrations of freeze-dried GelMA (7%, 15% (w/t)) were dissolved. Eosin Y sodium salt with 0.05 mM concentration was prepared separately in distilled water. The TEA/VC solution containing GelMA (7%, 15% (w/t)) was mixed with Eosin Y, before crosslinking to form the final precursor solution. The precursor solution prepared to form hydrogel was pipetted at 1 mm intervals between two glass slides and exposed to visible (VL) for 120 s, at a distance of 5 cm. The GelMA/BCP-NP mixture was prepared as a composite hydrogel, through the same process as above. Briefly, BCP-NP was added to the GelMA aqueous solution containing TEA/VC. After that, ultrasonic treatment was performed for 2 h, to uniformly disperse BCP-NP in the solution. The dispersed solution was mixed with Eosin Y, to carry out the photopolymerization. The final precursor solution formed polymerization by visible light, for 120 s, to form a composite hydrogel. The resulting hydrogel was labeled GelMA-XX, according to the concentration of GelMA. BCP is labeled as GelMA-XXBYY, according to its weight (example: GelMA-7, GelMA15, GelMA15B0.1, and GelMA15B10). The prepared samples were stored in a refrigerator, at −20 °C, until use. A schematic diagram of the synthesis of GelMA, based on the reaction of gelatin with methacrylic anhydride and photo-initiator, is shown in Figure 1.

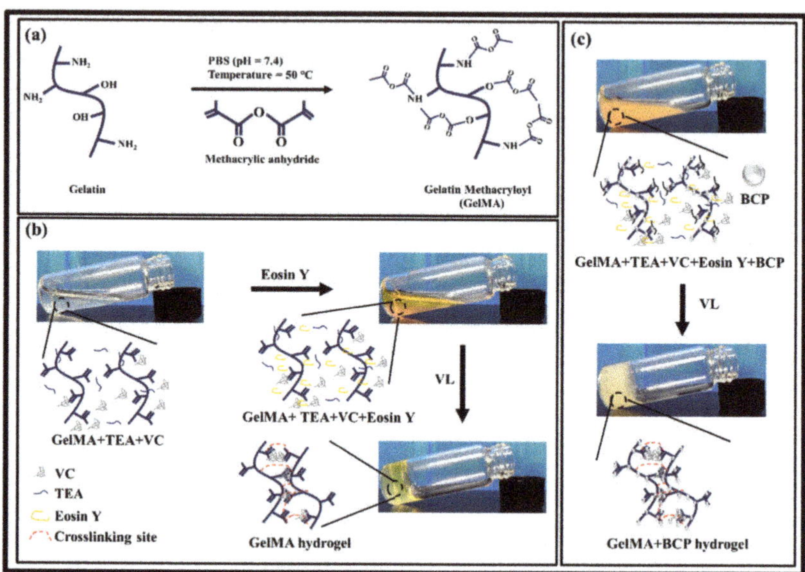

Figure 1. Schematic diagram of network structure formation on (**a**) gelatin methacryloyl, (**b**) gelatin methacryloyl hydrogel, and (**c**) biphasic calcium phosphate nanoparticles (BCP-NPs)-bonded gelatin methacryloyl hydrogel.

2.5. Characterization of BCP-NP and Composite Hydrogel

To confirm the crystal phase of samples were freeze-dried for 1 day and cut into the appropriate sizes, for analysis, by FE-SEM, FTIR, and XRD. FTIR analysis was performed within the wavelength 4000–500 cm^{-1} (KBr), using the attenuated total reflectance (ATR) method. BCP-NP and fabricated hydrogels were analyzed by XRD in a 20–60° range, with a 2° per minute gap at 60 kV. For FE-SEM, samples were sputter-coated with platinum and measured under an acceleration voltage of 10 kV.

2.6. Mechanical Analysis of GelMA Composite Hydrogel

For compression tests, GelMA and GelMA composite hydrogels of varying compositions were prepared as cylindrical samples with a 15 mm diameter and 2 mm thickness, which were measured by using a universal testing machine (Instron 5569). Hydrogels used for the measurement were hydrated in deionized water, for 24 h, and the experiment was performed at the maximum hydration condition. For accurate measurement, the sample was fixed between two parallel plates, and a 500 N load cell was mounted until fracture at a speed of 0.5 mm/min. The measured data were obtained, using Bluehill 2 software, and the modulus of elasticity was calculated at the first 10% strain of the curve in the stress–strain curve.

2.7. Hydrogel Degradation Rate Evaluation

The evaluation of the degradation rate of the prepared hydrogel and composite hydrogel was performed according to the following method [11]. Then, 25 U/mL^{-1} of collagenase (TypeII, Worthington Biochemical Co.) was prepared in PBS, 0.5 mL each was added to the hydrogel sample, and then gently shaken on a shaker, at 37 °C. After that, the hydrogel samples were taken out at various times, and the dried weight was measured. The results are expressed as $(M_t/M_0)^{1/2}$, where M_0 is the initial weight of the hydrogel, and M_t is the dry weight measured at time, t.

2.8. Thermogravimetric Analysis of Hydrogel (TGA)

GelMA and composite hydrogels were used for TGA. Hydrogel was freeze-dried after freezing at $-80\ °C$. The thermal behavior of the samples was carried out on an SDT Q 600 unit, with a $10\ °C/min$ heating lamp, from 25 to 600 $°C$.

2.9. Cell Viability

Cell experiments were conducted in this study by using the MC3T3-E1 preosteoblast that was provided by ATCC (American Type Culture Collection). For the culture medium, 10% FBS of a nutrient component containing antibiotics was added to an α-MEM medium. Cell culture was performed in an incubator (3111, Thermo Electron Corporation, Waltham, MA, USA), at 37 $°C$, in a 5% CO_2 atmosphere.

2.9.1. Water-Soluble Tetrazolium Salt (WST) Assay

The proliferation capacity of MC3T3-E1 cells against the prepared composite hydrogel was measured, using a water-soluble tetrazolium salt (WST-8) assay (ALX-850, Enzo Life Sciences, Farmingdale, NY, USA). After a hydrogel is swelled to the maximum hydration, it was cut to a 10 mm diameter and placed in a glass bottle, followed by sterilization by autoclaving at 121 $°C$ for 20 min. The sterilized sample was immersed in the medium for 1 h and then placed on the 12-well plate. The MC3T3-E1 cells were cultured for 3 and 5 days, at a cell density of 1×10^5 cells mL^{-1}. After 3 and 5 days, the medium was removed, and a 400 μL mixed solution containing water-soluble tetrazolium salt reagent and the α-MEM medium was dispensed and then stored in the 5% CO_2 incubator. After 90 min, 100 μL was added to a 96-well plate, and the absorbance was measured at 450 nm, using the ELISA reader (Molecular devices, Emax, San Jose, CA, USA).

2.9.2. Alkaline Phosphatase (ALP) Activity

ALP activity was evaluated, using the TRACP & ALP assay kit. The TRACP & ALP assay kit can detect the activity of acidic phosphatase (ACP) and alkaline phosphatase (ALP), respectively. In this study, it was used to detect ALP, an enzyme marker in osteoblasts. The cell culture method is the same as the method used in the WST assay. After 7 days of incubation, the differentiation of the preosteoblast cells was evaluated through the expression of ALP activity. The plates and composite hydrogels were washed, using saline. After that, the P-nitro-phenyl phosphate (pNPP) solution containing the ALP buffer solution and the extraction solution was added, following the kit protocol, and was allowed to react in the incubator, at 37 $°C$, for 1 h. After reaction completion, the ALP activity was evaluated by dispensing 100 μL of the solution in a 96-well plate and measuring the absorbance at 405 nm, using the ELISA reader.

2.10. Statistical Processing

Statistical analysis for all experimental results was conducted by using one-way analysis of variance with the Tukey test. A p-value lower than 0.05 was considered statistically significant (* $p < 0.05$, NS $p > 0.05$).

3. Results

3.1. Synthesis and Characterization of GelMA Macromers

The 1H-NMR spectra of gelatin and gelatin methacryloyl are shown in Supplementary Materials Figure S1. Compared with gelatin, derivatization by methacryloyl reaction was confirmed in the GelMA. In methacrylic anhydride, the shift of methacrylate protons was found at 1.8 ppm. In GelMA, a decrease in signal was observed at 2.9 ppm, due to the reaction of lysine amino acid and methacrylic acid. In addition, an increase in 5.4 and

5.7 ppm signals of gelatin methacryloyl was observed by the integrated lysine amino acid. The degree of methacrylate substitution (DOS) of GelMA was calculated by comparing the 2.9 ppm protons of unmodified gelatin and modified gelatin, and the calculated equation is as below (1) [23].

$$DOS = 1 - (\text{lysine methylene proton of GelMA})/(\text{lysine methylene proton of gelatin}) \times 100\% \tag{1}$$

3.2. Chemical Properties

The FTIR spectra of hydroxyapatite (HAP) and Tri-calcium phosphate (TCP) for comparison between gelatin hydrogel and composite hydrogel with BCP-NP and BCP-NP are shown in Figure 2a,b. In GelMA-15, the characteristic bands of hydrogels, such as NH amino band (1550 and 1650 cm^{-1}), OH band (3200–3500 cm^{-1}) and carbonate band (1425 and 1450 cm^{-1}) were observed. In GelMA-15B10, it was confirmed that the intensity of the O-H band and N-H amino band decreased due to the incorporation of BCP-NP. In addition, an increase in the 1030 cm^{-1} band, which is a strong band assigned to the stretching of PO for HPO$_4$$^{2-}$ and PO$_4$$^{3-}$, and a newly generated 890 cm^{-1} band was shown. Compared to the HAP and TCP graphs, the graphs of different aspects could be confirmed in BCP-NP. A single phosphate band between 500 and 600 cm^{-1} is observed in HAP and TCP, but shows two well-decomposed phosphate bands in BCP-NP. In addition, only the BCP-NP was able to observe the band allocated to the stretching of PO (890 cm^{-1}) and the characteristic band (1343 cm^{-1}) caused by the bending of POH. The results of the XRD analysis for the hydrogel and BCP-NP are shown in Figure 2c,d. It was confirmed that the peak of the prepared BCP-NP coincided with the peak of the previous paper, and a mixed peak of HAP (25.9°, 31.8°, 46.7°, and 49.5°) and β-TCP (25.9°, 33.6°, and 53.2°) was observed [22]. The peak of BCP-NP was not observed in GelMA-15; only a hydrogel peak at a certain level was observed. In GelMA-15B10, both the peak of BCP-NP and the peak of hydrogel were included. In particular, all HAPs showing high peak values were observed, and no peaks of β-TCP, which are low peaks, were observed. In comparison between nanoparticles, most similar peak values were observed in BCP-NP, HAP, and TCP. However, the HAP and TCP (about 27° and 29°) peaks divided into two were merged into one, in the BCP-NP, showing a β-TCP peak, and it was confirmed that the HAP and TCP peaks observed between 34° and 35° disappeared in the BCP-NP.

3.3. Morphological Analysis

The surface shape of the composite hydrogel containing BCP-NP and GelMA-15 is shown in Figure 3. The surface of GelMA-15 contracted by light polymerization did not show any pore shape and has an irregular structure in the form of a spider web (Figure 3a,b). GelMA-15B0.1 shows some regular surface shape, due to electrostatic attraction between hydrogel and BCP-NP due to the application of a small amount of BCP-NP (Figure 3c,d). In GelMA-15B1, it was possible to confirm the appearance that an appropriate amount of BCP-NP was covering the surface (Figure 3e,f). In addition, the BCP-NP was not evenly distributed on the surface and was able to confirm the sporadically clustered form. GelMA-15B10 was able to confirm that the entire surface was covered as an excess of BCP-NP was applied, and a more even surface was confirmed than for GelMA-15B1 (Figure 3g,h). EDX results for each hydrogel group are presented in Supplementary Materials Figure S2.

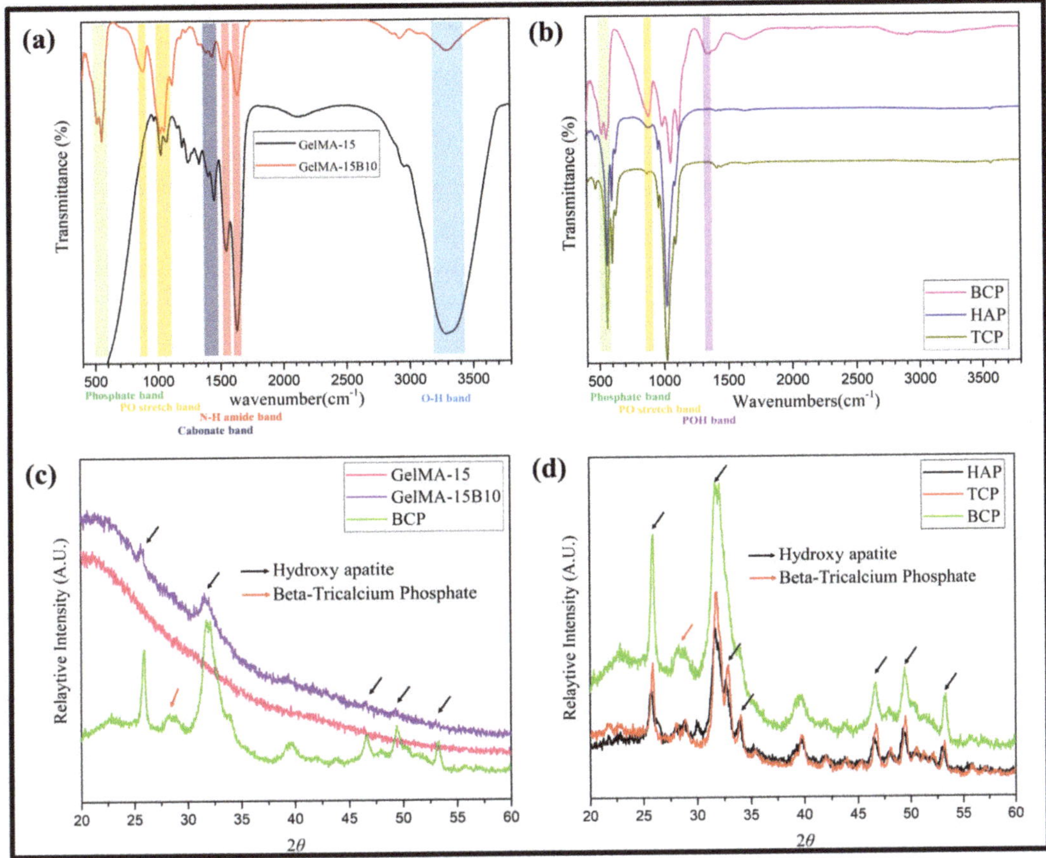

Figure 2. FTIR and XRD analysis of hydrogels (GelMA-15, GelMA-15B10, BCP-NP) and Nano-particles (BCP-NP, HAP, and TCP); FTIR; (**a**) hydrogels, (**b**) nano-particles/XRD, (**c**) hydrogels, and (**d**) nanoparticles.

3.4. Mechanical Properties

It is known that a composite hydrogel containing BCP-NP can easily control its mechanical properties, according to the weight of BCP-NP [24]. For confirmation, a mechanical strength test was performed by varying the weight of BCP-NP (0.1, 1, and 10 mg). The stress–strain graph obtained through the compression test is shown in Figure 4a. GelMA-15B0.1 showed no significant difference between GelMA-15 and maximum stress. However, it was found that the maximum stress of GelMA-15B1 was greatly improved, and that of GelMA-15B10 was about twice that of GelMA-15B1 and about 17 times that of GelMA-15. The compression modulus of the composite hydrogel containing BCP-NP and GelMA-15 is shown in Figure 4b. The compressive modulus was determined by the slope of the elastic region of the stress–strain curve. Like the stress–strain curve, as the ratio of BCP-NP increased, the compression ratio (4, 6, 81, and 93 kPa) was shown, and statistically significant differences were shown in all groups ($p < 0.05$). In particular, GelMA-15B1 showed a large increase of about 20 times, compared to GelMA-15, and GelMA-15B10 showed a difference of about 23 times. In addition, this result was shown to be similar to the previous stress–strain data, and the data for each group are shown in Supplementary Materials Table S1.

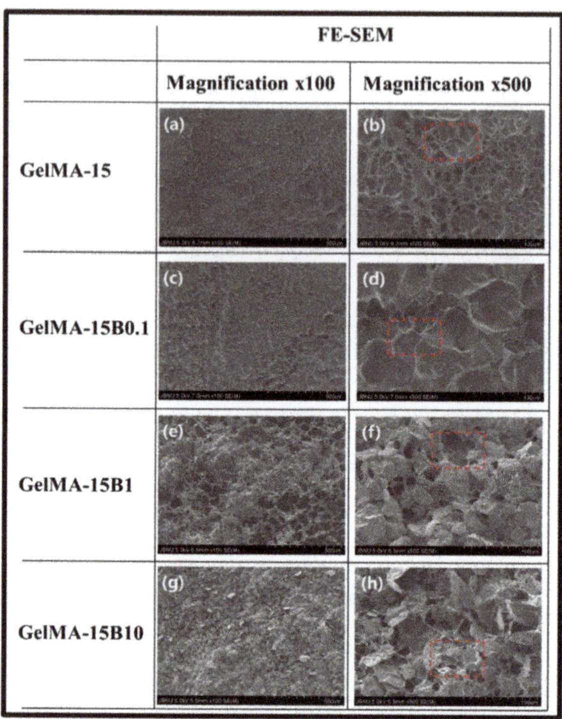

Figure 3. FE-SEM images on surface of (**a**,**b**) GelMA-15 and (**c**,**d**) GelMA-15B0.1, (**e**,**f**) GelMA-15B1 and (**g**,**h**) GelMA-15B10 hydrogels, at different magnifications ($\times100$ and $\times500$). Red area: surface morphology according to BCP capacity increase.

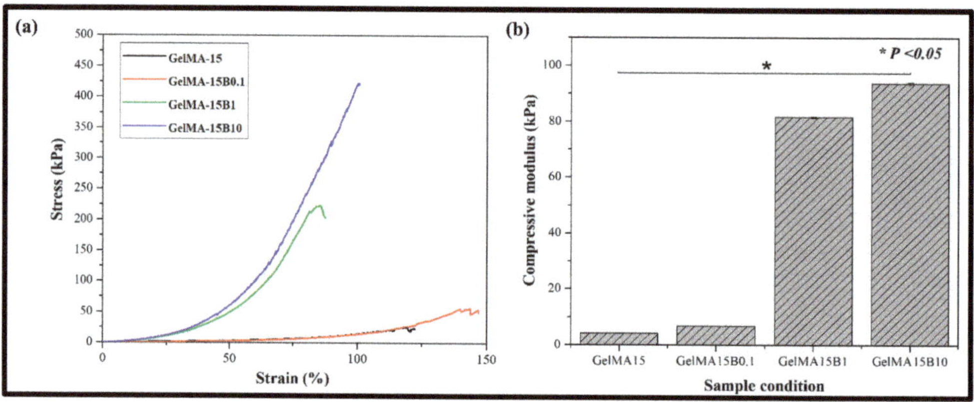

Figure 4. Graphs of (**a**) compressive stress–strain curve and (**b**) compressive modulus on the GelMA-15, GelMA-15B0.1, GelMA-15B1, and GelMA-15B10 groups.

3.5. Thermal Properties

The thermal properties of the hydrogel were carried out through a thermal analyzer. The TGA curve shows that the hydrogel began to lose integrity between 50 and 450 and the mass decreased (Figure 5a). In particular, the hydrogel containing BCP-NP showed a reduced mass loss from 350 °C, under the influence of BCP-NP. The final mass loss of all hydrogels was identified as −87% for GelMA-15, −85% for GelMA-15B0.1, −84% for GelMA-15B1, and −80% for GelMA-15B10, respectively. It was confirmed that a large amount of weight loss occurred through the DTG curve (Figure 5b). The first degradation of the hydrogel started between about 50 and 100 °C, and the composite hydrogel containing BCP-NP showed a greater loss than the single hydrogel. However, at 200 to 417 °C, where large weight loss occurs, the composite hydrogel containing BCP-NP showed lower weight loss than GelMA-15. For example, the red square in Figure 5b shows that the composite hydrogel containing BCP-NP has lower mass loss than GelMA-15. The DSC curve showed several endothermic peaks and occurred largely between 50 and 100 °C. (Figure 5c). In particular, the endothermic peak of the composite hydrogel containing BCP-NP moved to a lower temperature than GelMA-15 (90 °C). For example, GelMA-15B10 showed a peak at 75 °C, due to an increase in the BCP-NP content in the hydrogel.

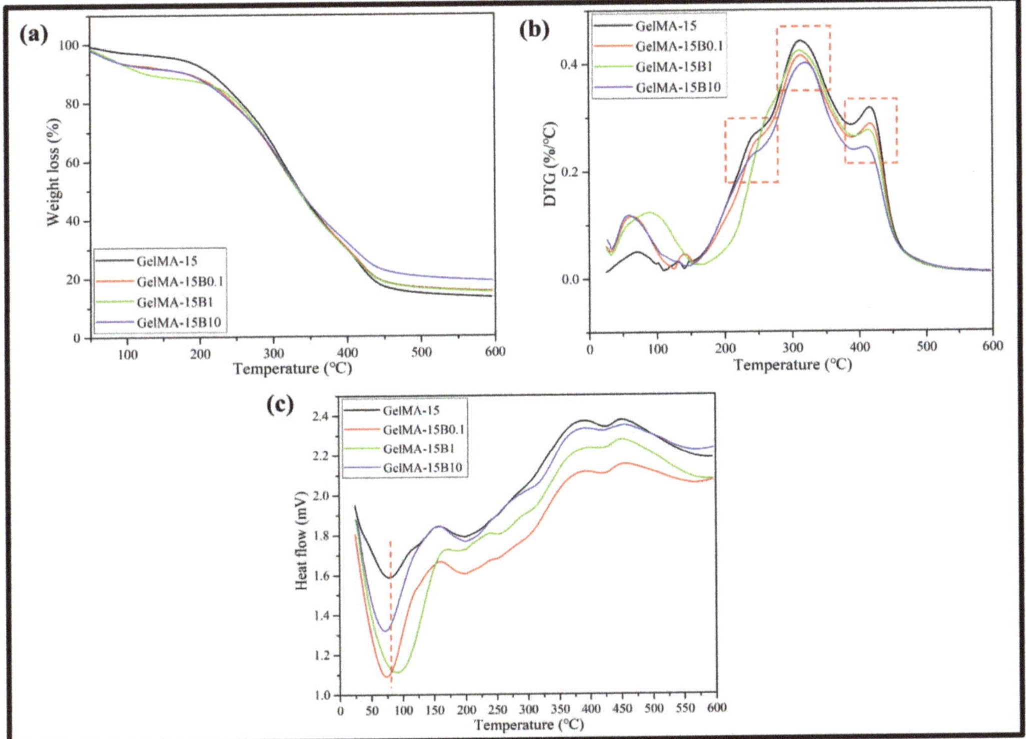

Figure 5. Thermal properties analysis of GelMA-15, GelMA-15B0.1, GelMA-15B1, and GelMA-15B10 hydrogels. (**a**) Thermogravimetric (TG) and (**b**) derived thermogravimetric (DTG) analysis. (**c**) Differential scanning calorimetry (DSC) curve.

3.6. Enzymatic Degradation

Hydrogels using gelatin have been confirmed to be biodegraded by enzymes, because gelatin contains a functional sequence that is stimulated by collagenase [8,25]. Therefore, to investigate the biodegradation effect of the composite hydrogel applied with GelMA and BCP-NP for enzymatic degradation, it was treated with collagenase type II. The remaining hydrogel was measured at a predetermined time (2, 4, 6, 8, 10, 15, 24, and 48 h). Figure 6 shows a graph of $(M_t/M_0)^{0.5}$ versus t. Moreover, $(M_t/M_0)^{0.5}$ represents the amount of degradation for the hydrogel, at the specified time, t, for the hydrogel and composite hydrogel. There was no difference in the degradation rate in all hydrogels, until the initial 2 h. However, from 8 h, GelMA-15 and hydrogel containing BCP-NP showed a large difference in degradation rate. In addition, GelMA-15, GelMA-15B0.1, and GelMA-15B1 groups were all degraded based on 15 h, and only the GelMA-15B10 group showed stability against enzymes for up to 48 h. The breakdown of the hydrogel is affected by the enzymatic cleavage of the GelMA gelatin backbone. However, the application of BCP-NP showed that the binding of the GelMA network and the BCP-NP could delay the enzyme cleavage by linking the chains of GelMA.

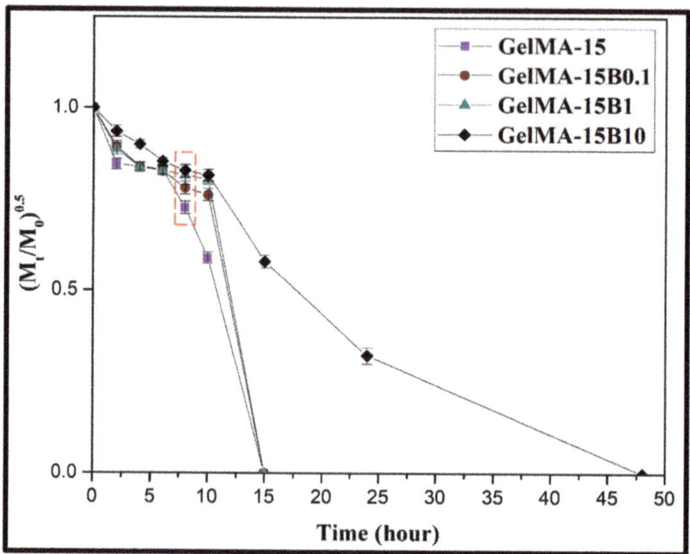

Figure 6. Biodegradation characteristics of GelMA-15, GelMA-15B0.1, GelMA-15B1, and GelMA15-B10 induced by treatment with collagenase type II (25 U/mL^{-1}).

3.7. Cell Viability against Preosteoblasts

Various concentrations of GelMA (7% and 15%) and GelMA-15B10 were selected for cell activity evaluation, and a control group in which only cells were dispensed was added for comparison. The results of cell proliferation were expressed in %, based on the control group (Figure 7a). As a result, during the initial three and five days, all groups improved more than 150%, as compared to the control group. However, there was no significant difference in GelMA-7, GelMA-15, and GelMA-15B10 (p >0.05). Cellular differentiation was confirmed through the expression of ALP activity (Figure 7b). Similar to the previous cell proliferation results, after seven days of culture, all hydrogel groups showed higher ALP activity than the control group, and showed a statistically significant difference (p <0.05). When comparing the hydrogel groups, GelMA-7 and GelMA-15B10 showed significantly increased ALP activity, as compared to GelMA-15 (p <0.05). However, there was no significant difference between GelMA-7 and GelMA-15B10 (p >0.05).

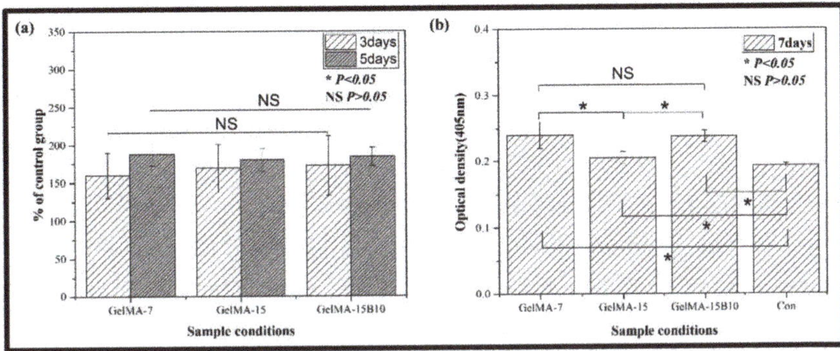

Figure 7. Proliferation (by water-soluble tetrazolium salt (WST)) and differentiation assessment (by alkaline phosphatase (ALP)) of MC3T3-E1 cells on the GelMA-7, GelMA-15, GelMA-15B10, and Control(con) groups after (**a**) three and five days of culture for WST assay and (**b**) seven days of culture for ALP assay.

4. Discussion

Synthesis of the substituted GelMA monomer by combining gelatin with MA confirmed the presence of methacrylate bound to the gelatin skeleton through 1H-NMR analysis. In Figure S1, no acrylic protons (6.1 ppm) were identified, and the discovery of GelMA's typical methacrylate proton C (5.4 ppm) and proton B (5.7 ppm) shows that MA was successfully grafted onto the gelatin skeleton. In addition, it was confirmed that the increase of 1.8 ppm and the decrease of 2.9 ppm, which are reactions of lysine protons in GelMA, were produced when $COOH^-$ of MA was combined with NH^- of gelatin [26,27].

The photo-crosslinkable hydrogel to which BCP-NP is physically bound was prepared, using a photo-initiator (TEA/VC/Eosin Y) mediating visible light. When exposed to UV rays, cells themselves and their functions are impaired, which may lead to tumors or cancers, so we tried to exclude them from the study [12,28]. Irgacure-2959 could not be activated with visible light because of its low water solubility and limited molar absorption. Although LAP has high water solubility and cellular compatibility, it has low activity in the visible range (405 nm) and high activity in the ultraviolet range (365–385 nm), making it unsuitable for use as an activation initiator in the visible range [29]. In particular, considering the effective wavelength (420–480 nm) of a dental-curing device that can be used clinically in tissue engineering, the possibility of using a cutting type photo-initiator is limited. To solve this problem, Eosin Y, known as an uncut photo-initiator, was used. Eosin Y not only minimizes the safety issues related to UV light, but can also be quickly activated with the effective wavelength used in dental-treatment systems. In addition, TEA and VC were used as co-initiators and monomers, respectively, to aid in free radical photoinitiation [30].

FTIR is known to be the most practical method for comparing organic compounds. The produced hydrogel and nanoparticles were further confirmed through FTIR and XRD analysis. The composite hydrogel was compared with GelMA-15 by selecting GelMA-15B10. The comparison of GelMA-15 and GelMA-15B10 in FTIR was confirmed through the fingerprint area. The fingerprint area is allocated in the area of 500~1500 cm^{-1} and is useful for comparison because it shows a unique pattern for organic compounds [31]. The binding of BCP-NP could be confirmed through the newly created PO stretching band (890 cm^{-1}) and phosphate band (500–600 cm^{-1}) (Figure 2a). Moreover, to confirm the generation of BCP-NP, comparison with HAP and TCP was conducted. BCP-NP and other nanoparticles also showed different properties (Figure 2b). HAP and TCP showed similar bands, but it was confirmed that only BCP-NP exhibited characteristic bands related to PO and POH [32], which confirmed that the production of BCP-NP was successful. GelMA-15B0.1 showed a wrinkled shape, instead of a hole shape, by mixing a little BCP-NP. However, as a result of observing the surface by using FE-SEM, the presence of BCP-NP could not be confirmed (Figure 3c,d).

The ability to maintain shape and mechanical properties during the material replacement and bone regeneration processes is the most ideal feature of guided bone generation and bone-graft materials. Hence, it is essential to maintain a balance between the degree of material degradation and bone formation. Achievement of improper mechanical strength results in the inability of the material to maintain its shape. Mimicking bone-forming tissues is an important strategy used to prepare biomimetic hydrogels for bone-tissue engineering. Approximately 35% of natural bone is made of organic substances (mainly type 1 collagen), while approximately 65% consists of inorganic substances (nanocrystalline calcium phosphate, HAP, and TCP) [33]. The composite hydrogel group containing BCP-NP showed higher stress, compared to the single hydrogel (Figure 4a). This seems to be because BCP-NP is bound to the hydrogel and acts as a crosslink to increase the strength of the material. The compressive elastic modulus also showed a tendency to increase as the ratio of BCP-NP increased. These results showed a tendency to agree with the previous stress–strain curve. In addition, the increase in the elasticity of the hydrogel can be an index, showing that it can resist external pressure when implanted at a defect site.

The thermal behavior of hydrogel samples for various thermal applications was characterized based on TGA (Figure 5). From the results of TGA and DTG, all hydrogels showed some mass loss up to 100 °C, which corresponds to dehydration, the thermal behavior of GelMA-based hydrogels. It is known that the mass loss of the hydrogel starting at 200 °C (Figure 5a) is mainly due to hydrogen bonds between molecules and the breakdown of molecular side chains of the gel component [34]. The observed thermal behavior of the hydrogel is due to previously published performance, and the addition of BCP nanoparticles to the polymer matrix did not significantly alter the polymer's thermal properties. This indicates a large weight loss of the entire sample because the decomposition process of GelMA-15 has already ended at about 375 °C, and in the case of the composite hydrogel containing BCP-NP, the degradation process ended at the same temperature as GelMA-15. However, the BCP-NP integration was confirmed due to the result of the remaining mass loss of the sample. Bone-regeneration conditions are determined by the interaction between the cells at the site of the bone defect and the surface of the implant material. When the composite hydrogel is implanted, the immune system will recognize the antigenic determinants provided to the hydrogel, and the immune system will probably reject the transplant [35]. Thus, one of the success factors for hydrogel application is to reduce immune rejection and increase the biocompatibility of biomaterials. The cell compatibility of the hydrogel was confirmed by using mouse-derived preosteoblasts (MC3T3-E1 cells) and cell proliferation and differentiation tests (Figure 7a,b). The higher the concentration of GelMA, the stronger the mechanical properties, but the lower the cell activity. In the previous literature, it was confirmed that a hydrogel with a GelMA concentration of 7–8% exhibits high cellular activity [36]. In this study, a hydrogel with a GelMA concentration of 15% was prepared, and a GelMA concentration of 7% hydrogel was additionally prepared to compare cell activity. Cell proliferation was not significantly different in all hydrogel groups but increased significantly compared to the control group. However, the differentiation of cells by ALP activity showed different results. As reported in the previous literature, GelMA-7 showed better cell differentiation than GelMA-15, and a statistically significant difference was also confirmed. However, GelMA-15B10 showed better cell differentiation than GelMA-15, and there was no significant difference from GelMA-7. It can be seen that BCP-NP contributed to increased cell differentiation. Bone tissue has a hierarchical structure ranging from macroscopic to nanoscale, and many studies have attempted to mimic it. It is also known that hydrogels provide a dynamic environment similar to that of the extracellular matrix [37,38]. In this study, nanosized BCP-NP was contained in a composite hydrogel, and a nano-shaped surface was provided for cell adhesion. BCP-NP is known to stimulate cellular activity by stimulating the proliferation of hMSCs [39] but is expected to sufficiently affect preosteoblasts (MC3T3-E1) and promote cellular activity.

5. Conclusions

In this study, a hydrogel applicable to bone-tissue engineering was fabricated, using gelatin methacrylate with excellent biocompatibility, and characteristics were evaluated by applying BCP-NP. According to the analysis, the nano BCP-NP is a mixed form of HA and b-TCP, and characteristic spectra and peaks of BCP-NP were confirmed through XRD and FTIR. The composite hydrogel (GelMA-15B10) improved mechanical properties and increased up to 23 times more than GelMA-15. In addition, through enzymatic degradable analysis, it was confirmed that the composite hydrogel has better durability than the single hydrogel. The visible light initiator and BCP-NP used in the hydrogel production showed good results for preosteoblast cells (MC3T3-E1), and, in particular, BCP-NP improved cell differentiation in the hydrogel. Based on these results, the composite hydrogel to which BCP-NP is applied not only has stability against biodegradation but is expected to be used as a material applicable to the field of bone-tissue engineering in the future.

Supplementary Materials: The following are available online at https://www.mdpi.com/2079-499 1/11/3/617/s1, Figure S1. 1H-NMR spectra of gelatin and gelatin methacryloyl (GelMA) monomer with degree of substitution; Figure S2. The energy-dispersive X-ray (EDX) profile of (a) GelMA-15, (b) GelMA-15B0.1, (c) GelMA-15B1, (d) GelMA-15B10.; Table S1. Maximum stress and Compressive Modulus for each hydrogel groups.

Author Contributions: Conceptualization and methodology, J.-B.C. and Y.-K.K.; software, J.-E.P.; validation and formal analysis, Y.-K.K. and J.-E.P.; investigation and resources, T.-S.B.; data curation, S.-M.B.; writing—original draft preparation, J.-B.C.; writing—review and editing, J.-B.C. and Y.-S.J.; visualization, S.-M.B.; project administration, Y.-K.K.; supervision and funding acquisition, M.-H.L. All authors have read and agreed to the published version of the manuscript.

Funding: This research was funded by a National Research Foundation of Korea (NRF) grant funded by the Korea Government (MSIP), No. 2019R1C1C1003784.

Data Availability Statement: The data presented in this study are available in Supplementary Material file.

Acknowledgments: This study has reconstructed the data of Ji-Bong Choi's doctoral dissertation submitted in 2021 (title: "Fabrication and characterization of biodegradable gelatin-methacrylate hydrogel for tissue engineering").

Conflicts of Interest: The authors declare no conflict of interest.

References

1. ScSculean, A.; Nikolidakis, D.; Nikou, G.; Ivanovic, A.; Chapple, I.L.; Stavropoulos, A. Biomaterials for promoting periodontal regeneration in human intrabony defects: A systematic review. *Periodontology 2000* **2015**, *68*, 182–216. [CrossRef] [PubMed]
2. Kao, R.T.; Nares, S.; Reynolds, M.A. Periodontal regeneration–Intrabony defects: A systematic review from the AAP regeneration workshop. *J. Periodontol.* **2015**, *86*, S77–S104. [CrossRef]
3. Athirasala, A.; Tahayeri, A.; Thrivikraman, G.; França, C.M.; Monteiro, N.; Tran, V.; Ferracane, J.; Bertassoni, L.E. A dentin-derived hydrogel bioink for 3D bioprinting of cell laden scaffolds for regenerative dentistry. *Biofabrication* **2018**, *10*, 024101. [CrossRef] [PubMed]
4. Monteiro, N.; Thrivikraman, G.; Athirasala, A.; Tahayeri, A.; França, C.M.; Ferracane, J.L.; Bertassoni, L.E. Photopolymerization of cell-laden gelatin methacryloyl hydrogels using a dental curing light for regenerative dentistry. *Dent. Mater.* **2018**, *34*, 389–399. [CrossRef] [PubMed]
5. Nichol, J.W.; Koshy, S.T.; Bae, H.; Hwang, C.M.; Yamanlar, S.; Khademhosseini, A. Cell-laden microengineered gelatin methacrylate hydrogels. *Biomaterials* **2010**, *31*, 5536–5544. [CrossRef] [PubMed]
6. Lin, R.-Z.; Chen, Y.-C.; Moreno-Luna, R.; Khademhosseini, A.; Melero-Martin, J.M. Transdermal regulation of vascular network bioengineering using a photopolymerizable methacrylated gelatin hydrogel. *Biomaterials* **2013**, *34*, 6785–6796. [CrossRef]
7. Fang, X.; Xie, J.; Zhong, L.; Li, J.; Rong, D.; Li, X.; Ouyang, J. Biomimetic gelatin methacrylamide hydrogel scaffolds for bone tissue engineering. *J. Mater. Chem. B* **2016**, *4*, 1070–1080. [CrossRef]
8. Xiao, S.; Zhao, T.; Wang, J.; Wang, C.; Du, J.; Ying, L.; Lin, J.; Zhang, C.; Hu, W.; Wang, L. Gelatin methacrylate (GelMA)-based hydrogels for cell transplantation: An effective strategy for tissue engineering. *Stem Cell Rev. Rep.* **2019**, *15*, 664–679. [CrossRef]

9.	Luo, Z.; Sun, W.; Fang, J.; Lee, K.; Li, S.; Gu, Z.; Dokmeci, M.R.; Khademhosseini, A. Biodegradable gelatin methacryloyl microneedles for transdermal drug delivery. *Adv. Healthc. Mater.* **2019**, *8*, 1801054. [CrossRef]

10.	Eslami, M.; Vrana, N.E.; Zorlutuna, P.; Sant, S.; Jung, S.; Masoumi, N.; Khavari-Nejad, R.A.; Javadi, G.; Khademhosseini, A. Fiber-reinforced hydrogel scaffolds for heart valve tissue engineering. *J. Biomater. Appl.* **2014**, *29*, 399–410. [CrossRef]

11.	Cha, C.; Shin, S.R.; Gao, X.; Annabi, N.; Dokmeci, M.R.; Tang, X.; Khademhosseini, A. Controlling mechanical properties of cell-laden hydrogels by covalent incorporation of graphene oxide. *Small* **2014**, *10*, 514–523. [CrossRef]

12.	Wang, Z.; Abdulla, R.; Parker, B.; Samanipour, R.; Ghosh, S.; Kim, K. A simple and high-resolution stereolithography-based 3D bioprinting system using visible light crosslinkable bioinks. *Biofabrication* **2015**, *7*, 045009. [CrossRef] [PubMed]

13.	Cidonio, G.; Alcala-Orozco, C.R.; Lim, K.S.; Glinka, M.; Mutreja, I.; Kim, Y.-H.; Dawson, J.I.; Woodfield, T.B.; Oreffo, R.O. Osteogenic and angiogenic tissue formation in high fidelity nanocomposite Laponite-gelatin bioinks. *Biofabrication* **2019**, *11*, 035027. [CrossRef]

14.	Thakur, T.; Xavier, J.R.; Cross, L.; Jaiswal, M.K.; Mondragon, E.; Kaunas, R.; Gaharwar, A.K. Photocrosslinkable and elastomeric hydrogels for bone regeneration. *J. Biomed. Mater. Res. Part A* **2016**, *104*, 879–888. [CrossRef] [PubMed]

15.	Paknejad, M.; Emtiaz, S.; Rokn, A.; Islamy, B.; Safiri, A. Histologic and histomorphometric evaluation of two bone substitute materials for bone regeneration: An experimental study in sheep. *Implant Dent.* **2008**, *17*, 471–479. [CrossRef]

16.	García-Gareta, E.; Coathup, M.J.; Blunn, G.W. Osteoinduction of bone grafting materials for bone repair and regeneration. *Bone* **2015**, *81*, 112–121. [CrossRef] [PubMed]

17.	Giannoudis, P.V.; Dinopoulos, H.; Tsiridis, E. Bone substitutes: An update. *Injury* **2005**, *36*, S20–S27. [CrossRef]

18.	Lan, W.; Zhang, X.; Xu, M.; Zhao, L.; Huang, D.; Wei, X.; Chen, W. Carbon nanotube reinforced polyvinyl alcohol/biphasic calcium phosphate scaffold for bone tissue engineering. *RSC Adv.* **2019**, *9*, 38998–39010. [CrossRef]

19.	Nie, L.; Wu, Q.; Long, H.; Hu, K.; Li, P.; Wang, C.; Sun, M.; Dong, J.; Wei, X.; Suo, J. Development of chitosan/gelatin hydrogels incorporation of biphasic calcium phosphate nanoparticles for bone tissue engineering. *J. Biomater. Sci. Polym. Ed.* **2019**, *30*, 1636–1657. [CrossRef]

20.	Faruq, O.; Kim, B.; Padalhin, A.R.; Lee, G.H.; Lee, B.-T. A hybrid composite system of biphasic calcium phosphate granules loaded with hyaluronic acid–Gelatin hydrogel for bone regeneration. *J. Biomater. Appl.* **2017**, *32*, 433–445. [CrossRef]

21.	Zhao, X.; Lang, Q.; Yildirimer, L.; Lin, Z.Y.; Cui, W.; Annabi, N.; Ng, K.W.; Dokmeci, M.R.; Ghaemmaghami, A.M.; Khademhosseini, A. Photocrosslinkable gelatin hydrogel for epidermal tissue engineering. *Adv. Healthc. Mater.* **2016**, *5*, 108–118. [CrossRef]

22.	Chen, Y.; Li, J.; Kawazoe, N.; Chen, G. Preparation of dexamethasone-loaded calcium phosphate nanoparticles for the osteogenic differentiation of human mesenchymal stem cells. *J. Mater. Chem. B* **2017**, *5*, 6801–6810. [CrossRef] [PubMed]

23.	Li, X.; Chen, S.; Li, J.; Wang, X.; Zhang, J.; Kawazoe, N.; Chen, G. 3D culture of chondrocytes in gelatin hydrogels with different stiffness. *Polymers* **2016**, *8*, 269. [CrossRef]

24.	Chen, Y.; Kawazoe, N.; Chen, G. Preparation of dexamethasone-loaded biphasic calcium phosphate nanoparticles/collagen porous composite scaffolds for bone tissue engineering. *Acta Biomater.* **2018**, *67*, 341–353. [CrossRef]

25.	van den Steen, P.E.; Dubois, B.; Nelissen, I.; Rudd, P.M.; Dwek, R.A.; Opdenakker, G. Biochemistry and molecular biology of gelatinase B or matrix metalloproteinase-9 (MMP-9). *Crit. Rev. Biochem. Mol. Biol.* **2002**, *37*, 375–536. [CrossRef]

26.	Hu, X.; Ma, L.; Wang, C.; Gao, C. Gelatin hydrogel prepared by photo-initiated polymerization and loaded with TGF-β1 for cartilage tissue engineering. *Macromol. Biosci.* **2009**, *9*, 1194–1201. [CrossRef]

27.	Claaßen, C.; Claaßen, M.H.; Truffault, V.; Sewald, L.; Tovar, G.n.E.; Borchers, K.; Southan, A. Quantification of substitution of gelatin methacryloyl: Best practice and current pitfalls. *Biomacromolecules* **2018**, *19*, 42–52. [CrossRef]

28.	Sinha, R.P.; Häder, D.-P. UV-induced DNA damage and repair: A review. *Photochem. Photobiol. Sci.* **2002**, *1*, 225–236. [CrossRef]

29.	Shih, H.; Lin, C.C. Visible-light-mediated thiol-Ene hydrogelation using eosin-Y as the only photoinitiator. *Macromol. Rapid Commun.* **2013**, *34*, 269–273. [CrossRef]

30.	Noshadi, I.; Hong, S.; Sullivan, K.E.; Sani, E.S.; Portillo-Lara, R.; Tamayol, A.; Shin, S.R.; Gao, A.E.; Stoppel, W.L.; Black, L.D., III. In vitro and in vivo analysis of visible light crosslinkable gelatin methacryloyl (GelMA) hydrogels. *Biomater. Sci.* **2017**, *5*, 2093–2105. [CrossRef]

31.	Choi, J.B.; Jang, Y.S.; Byeon, S.M.; Jang, J.H.; Kim, Y.K.; Bae, T.S.; Lee, M.H. Effect of composite coating with poly-dopamine/PCL on the corrosion resistance of magnesium. *Int. J. Polym. Mater. Polym. Biomater.* **2019**, *68*, 328–337. [CrossRef]

32.	Touny, A.; Saleh, M. Fabrication of biphasic calcium phosphates nanowhiskers by reflux approach. *Ceram. Int.* **2018**, *44*, 16543–16547. [CrossRef]

33.	Pina, S.; Oliveira, J.M.; Reis, R.L. Natural-based nanocomposites for bone tissue engineering and regenerative medicine: A review. *Adv. Mater.* **2015**, *27*, 1143–1169. [CrossRef]

34.	Liu, X.; Song, T.; Chang, M.; Meng, L.; Wang, X.; Sun, R.; Ren, J. Carbon nanotubes reinforced maleic anhydride-modified xylan-g-poly (N-isopropylacrylamide) hydrogel with multifunctional properties. *Materials* **2018**, *11*, 354. [CrossRef]

35.	Ramay, H.R.; Zhang, M. Biphasic calcium phosphate nanocomposite porous scaffolds for load-bearing bone tissue engineering. *Biomaterials* **2004**, *25*, 5171–5180. [CrossRef]

36.	Lee, D.; Choi, E.J.; Lee, S.E.; Kang, K.L.; Moon, H.-J.; Kim, H.J.; Youn, Y.H.; Heo, D.N.; Lee, S.J.; Nah, H. Injectable biodegradable gelatin-methacrylate/β-tricalcium phosphate composite for the repair of bone defects. *Chem. Eng. J.* **2019**, *365*, 30–39. [CrossRef]

37. Luo, Y.; Lode, A.; Wu, C.; Chang, J.; Gelinsky, M. Alginate/nanohydroxyapatite scaffolds with designed core/shell structures fabricated by 3D plotting and in situ mineralization for bone tissue engineering. *ACS Appl. Mater. Interfaces* **2015**, *7*, 6541–6549. [CrossRef] [PubMed]
38. Zhao, C.; Xia, L.; Zhai, D.; Zhang, N.; Liu, J.; Fang, B.; Chang, J.; Lin, K. Designing ordered micropatterned hydroxyapatite bioceramics to promote the growth and osteogenic differentiation of bone marrow stromal cells. *J. Mater. Chem. B* **2015**, *3*, 968–976. [CrossRef] [PubMed]
39. Chen, Y.; Wang, J.; Zhu, X.; Tang, Z.; Yang, X.; Tan, Y.; Fan, Y.; Zhang, X. Enhanced effect of β-tricalcium phosphate phase on neovascularization of porous calcium phosphate ceramics: In vitro and in vivo evidence. *Acta Biomater.* **2015**, *11*, 435–448. [CrossRef] [PubMed]

 *nanomaterials*

Article

Human Teeth-Derived Bioceramics for Improved Bone Regeneration

Ki-Taek Lim [1,*,†], Dinesh K. Patel [1,†], Sayan Deb Dutta [1], Han-Wool Choung [2], Hexiu Jin [3], Arjak Bhattacharjee [4] and Jong Hoon Chung [5,*]

[1] Department of Biosystems Engineering, Kangwon National University, Chuncheon 24341, Korea; dbhu10@gmail.com (D.K.P.); sayan91dutta@gmail.com (S.D.D.)

[2] Department of Oral and Maxillofacial Surgery and Dental Research Institute, School of Dentistry, Seoul National University, Seoul 151921, Korea; woolmania@naver.com

[3] Department of Plastic and Traumatic Surgery, School of Stomatology, Beijing Stomatological Hospital, Capital Medical University, Beijing 100069, China; hyuksoo725@daum.net

[4] Department of Materials Science and Engineering, Indian Institute of Technology, Kanpur 208016, India; arjak.ceramic16@gmail.com

[5] Department of Biosystems and Biomaterials Science and Engineering, Seoul National University, Seoul 151921, Korea

* Correspondence: ktlim@kangwon.ac.kr (K.-T.L.); jchung@snu.ac.kr (J.H.C.)

† These authors contributed equally to the paper.

Received: 3 November 2020; Accepted: 23 November 2020; Published: 30 November 2020

Abstract: Hydroxyapatite (HAp, $Ca_{10}(PO_4)_6(OH)_2$) is one of the most promising candidates of the calcium phosphate family, suitable for bone tissue regeneration due to its structural similarities with human hard tissues. However, the requirements of high purity and the non-availability of adequate synthetic techniques limit the application of synthetic HAp in bone tissue engineering. Herein, we developed and evaluated the bone regeneration potential of human teeth-derived bioceramics in mice's defective skulls. The developed bioceramics were analyzed by X-ray diffraction (XRD), Fourier-transform infrared (FTIR) spectroscopy, and scanning electron microscopy (FE-SEM). The developed bioceramics exhibited the characteristic peaks of HAp in FTIR and XRD patterns. The inductively coupled plasma mass spectrometry (ICP-MS) technique was applied to determine the Ca/P molar ratio in the developed bioceramics, and it was 1.67. Cytotoxicity of the simulated body fluid (SBF)-soaked bioceramics was evaluated by WST-1 assay in the presence of human alveolar bone marrow stem cells (hABMSCs). No adverse effects were observed in the presence of the developed bioceramics, indicating their biocompatibility. The cells adequately adhered to the bioceramics-treated media. Enhanced bone regeneration occurred in the presence of the developed bioceramics in the defected skulls of mice, and this potential was profoundly affected by the size of the developed bioceramics. The bioceramics-treated mice groups exhibited greater vascularization compared to control. Therefore, the developed bioceramics have the potential to be used as biomaterials for bone regeneration application.

Keywords: human tooth powder; bioceramics; biocompatibility; bone regeneration; vascularization

---

1. Introduction

Total hip replacement surgery is a massive burden globally due to its tremendous negative impact on the socioeconomic scenario. The total number of hip replacement surgeries in the United States is anticipated to increase more than 1.5 times (~174%) by 2030 [1]. These alarming statistics might force the healthcare community to develop an advanced hospital management model for many patients in a limited space [2] and develop alternate implant material for effective hip replacement surgery [3,4]. Various kinds of bone grafts are utilized to reconstruct bone defects caused by disease or trauma [5]. Autografts are considered a "gold standard" for the bone replacement application. The distinct advantage of histocompatibility without disease transfer risks makes the autografts ideal for bone repair [5,6]. However, their limited availability forces researchers to develop alternative and adequate bone substitutes. Black phosphorus (BP) consists of a single phosphorus element, a homolog to the inorganic constituents of natural bone, and can be applied to treat bone defects [7]. Calcium phosphate-based bioceramics, such as hydroxyapatite (HAp), β-tricalcium phosphate, etc., are considered suitable bone graft materials due to their structural and chemical similarity to human bone [8]. These materials are also used as a bioactive coating, bone cement, or in drug delivery applications. Bone contains nanometer-sized crystalline calcium phosphate with dimensions of ~5–20 nm width and 60 nm length [9,10]. Various reports are available highlighting the nano-bioceramic potential for bone tissue applications [11,12].

It has been observed that nanocrystalline HAp powders promote osteoblast adhesion, proliferation, and vascularization [13–16]. The promising features of nanocrystalline HAp, including better biocompatibility, lack of inflammatory, immunity reactions, and improved osteo-integrative potentials, have appealed to the researchers to fabricate nano-structured scaffolds that mimic natural bone properties [17]. Nanomaterials exhibit fascinating and superior mechanical, optical, and electronic properties due to their high surface-to-volume ratio and specific structural properties that do not occur in micro and/or macro-sized analogs [18]. A significant advancement in the fabrication and application of nanomaterials has been achieved. However, nano-calcium phosphate's practical applications are still far in the future due to the unknown nano-toxicity and lack of a "standardized" approach to monitor nanomaterial toxicity [19,20]. Various methods, like sol–gel synthesis, co-precipitation, mechanochemical synthesis, hydrothermal reaction, and microemulsion technique, have been applied for the synthesis of nanocrystalline HAp. Among these synthesis techniques, the synthesis quantities are significant concerns, and most of them are restricted to synthesis in small amounts [21]. Therefore, it is necessary to develop an alternative method for synthesizing nanomaterials from natural sources, such as mammalian bone, aquatic sources, shell sources, and mineral sources with improved biocompatibility [22].

We previously evaluated the effects of the sintering temperature for the formation of calcium phosphate-based bioceramics from human teeth for tissue engineering applications. The results indicated that the sintering temperature plays a crucial role in the characteristic properties of the developed bioceramics [23]. Herein, we developed the different sizes of bioceramics from human teeth through heat treatment and comparatively evaluated their bone regeneration potential with commercially available Bio-Oss® (Geistlich Pharma AG, Wolhusen, Switzerland), which is often applied as a bone implant. The developed bioceramics exhibited the Ca/P ratio of 1.67, close to the commercially available HAp. The biocompatibility of prepared samples was monitored by WST-1 assay in the presence of human alveolar bone-derived mesenchymal stem cells (hABMSCs). The cells were healthy and properly adhered to the developed bioceramics. Improved bone regeneration was observed in the defected skulls of mice with the prepared bioceramics, and this potential is profoundly affected by the size of the sample. Higher vascularization occurred in bioceramics-treated mice groups than the control, showing the developed material's potential for enhanced bone regeneration. The vascularization density was high nano-sized bioceramics compared to micro-sized and commercially available Bio-Oss® material. Therefore, the bone regeneration potential of the developed bioceramics can be fine-tuned by taking different particle sizes.

2. Methods

This study has four sections; (a) preparation, (b) analysis of the samples, (c) in vitro, and (d) in vivo investigations related to bone regeneration efficiency.

2.1. Preparation of Nano-Calcium Phosphate Bioceramic

Human teeth (18–35 years, Numbers-15) were received from the Department of Oral and Maxillofacial Surgery, Dental Hospital, Seoul National University, Republic of Korea. All procedures for teeth handling were performed under the regulation of an experimental protocol permitted by the Institutional Review Board (IRB) of the Dental Hospital, Seoul National University, Seoul, Korea (IRB No. CRI05008), and informed consent was obtained from each donor. Herein, we performed the 15 donors' pool to eliminate the donor-specific variability. The received teeth were washed with distilled water to remove the soft tissues from the surface and dried at 50 °C for 48 h in an air oven. The dried teeth were crushed into powders, and crushed powders were heated in an electric furnace (ST-01045, Daihan Scientific, Seoul, Korea) at 1000 °C with a heating rate of 10 °C/min for 2 h in air atmosphere. The heat-treated samples were pulverized with a miller (A10, IKA-WERKE, Nara, Japan), followed by the separation of the developed bioceramics by a sieve (Sieve/Shaker, Daihan Scientific). The diameter of the developed bioceramics occurred in the range of ~50–500 nm. The obtained bioceramics were treated and scrutinized with a NanoSizer Fine Mill (Deaga Powder Systems Co. Ltd., Seoul, Korea) to obtain the nano-sized bioceramics. The sterilization of the samples was performed in an autoclave, followed by cleaning with ultraviolet treatment. The simulated body fluid (SBF) was initially used to evaluate the biomineralization potential of the developed bioceramics ($n = 3$) for different periods (6 and 12 months) at 37 °C. The chemical compositions of the SBF are given in Table S1. The SBF solution was kept under mild conditions of pH and near to physiological temperature.

2.2. Phase and Microstructural Characterization of Nano-Calcium Phosphate Bioceramics

The microstructure and particle size of fresh and SBF-soaked bioceramics ($n = 3$) were analyzed by a field emission scanning electron microscope (FE-SEM, SUPRA 55VP, Carl Zeiss, Oberkochen, Germany). For this, the sample (fresh and SBF-soaked) powders were coated by a BAL-TEC SCD005 sputter coater for 250 s at 15 mA. The compositional analysis of the fresh and SBF-soaked bioceramics was performed by using energy-dispersive X-ray (EDX) spectroscopy at 30.0 kV. The phase analysis of the fresh bioceramics was accomplished by X-ray diffraction (XRD) (Bruker D5005 X-ray Diffractometer, Kassel, Germany) at room temperature using Cu Kα as the radiation source with a scan speed of 1°/min in the range of 10–90° and an angular range (2θ) (generator was 40 kV, 40 mA and λ (radiation) was 1.5406). The X-ray fluorescence (XRF) spectroscopy (Bruker S4 Pioneer, Karlsruhe, Germany) was used to determine the Ca/P ratio in fresh bioceramics and commercially available HAp using a Rh X-ray source and helium atmosphere. The Fourier-transform infrared (FTIR) spectroscopy (Nicolet 6700, Thermo Scientific, Madison, WI, USA) was performed to determine the functional groups present in fresh and SBF-soaked bioceramics in the range of 400–4000 cm^{-1}. The thermal stability of the fresh bioceramics was evaluated through thermal gravimetric analysis (TGA) (SDT Q600, TA Instruments, New Castle, DE, USA) with a 10°/min heating rate from room temperature to 1400 °C.

2.3. Cell Culture and Maintenance

The hABMSCs were received from the Korean Cell Line Bank (KCLB; Seoul National University, Korea), and cell culture was performed according to the method described by Gronthos et al. [24]. Briefly, the cells were cultured in α-minimum essential media (MEM) containing 10% fetal bovine serum (FBS, Welgene Inc., Gyeongsan, Korea), 10 mM ascorbic acid (L-ascorbic acid), 1% antibiotics (Anti-Anti, 100×, Gibco-BRL, Gaithersburg, MD, USA), and sodium bicarbonate (Sigma-Aldrich, St. Louis, MO, USA) at 37 °C in a humidified atmosphere with 5% CO$_2$ (Steri-Cycle 370 Incubator, Thermo Fisher Scientific, USA) for desired periods. Passage three hABMSCs were used for in vitro experiments.

2.4. Cytotoxicity Evaluation

The indirect cell viability of hABMSCs was evaluated by WST-1 assay (EZ-Cytox Cell Viability Assay Kit, Daeillab Service Co. Ltd., Seoul, Korea) in the presence of the developed bioceramics as reported earlier in somewhere else [25]. In brief, the extraction media were prepared by soaking the samples in a serum-free media (SFM), and incubated for 24 h. The cells (1×10^4) were seeded in 96-well plates in SFM and incubated at 37 °C and 5% CO_2 environment for 24 h. After this, the media were changed with the extracted media and further incubated for 24 h. The media without any extract were considered as control. After 24 h of incubation, the cultured cells were washed with PBS and treated with 10 μL of WST-1 dye and further incubated for 2 h. The formed formazan concentration was measured with a spectrophotometer (Victor 3, Perkin Elmer) at 450 nm (reference wavelength 625 nm). All experiments were performed in triplicate ($n = 3$, size ~60 nm), and values are expressed as mean ± standard deviation (SD).

2.5. Cell Morphology

The adhesion behavior of hABMSCs in the presence of the prepared bioceramics was evaluated through FE-SEM after 1 and 7 days of the treatment. Briefly, the cells were seeded in a 35 × 10 mm culture disc in the presence of the developed bioceramics ($n = 3$, size ~60 nm) and incubated at 37 °C and 5% CO_2 conditions. The cultured cells were washed with PBS and fixed with 4% paraformaldehyde (PFA) (Sigma-Aldrich, USA) for 1 h at 4 °C. The fixed cells were washed with PBS and treated with ethanol. The samples were sputter-coated with platinum, and images were captured by an FE-SEM.

2.6. Animal Care, In Vivo Bone Formation and Vascularization Study

The imprinting control region (ICR) (male, 32–34 g, six weeks old) mice were purchased from Orient Bio, Gapyeong-gun, Korea. All mice were kept in an insulated and soundproof room at an ambient temperature of 21 ± 2 °C, with a constant relative humidity of 35 ± 2% with an automatically controlled 12 h light and 12 h dark cycle (lights off at 20:00 h). Sufficient amounts of food and water were supplied to the mice for their care. The experimental mice were divided into two groups (Group = 2, and total mice = 6). The one group represented the negative (without any treatment) and positive control (with commercially available Bio-Oss®, size 0.25–1 mm, Geistlich Pharmaceutical, Wolhusen, Switzerland), and the other group showed the experimental conditions with different sizes of bioceramics. The in vivo study was performed as reported in an earlier work [23]. Briefly, the mice were treated with ether, followed by the intraperitoneal supplement of 0.3–0.4 mL of 20 mg/mL Avertin (Sigma-Aldrich, USA) for anesthetization. The 5% iodine solution was used to disinfect the mice skull. After this, a 1–1.5 cm sagittal incision was created very carefully on the scalp with a sterile surgical blade, followed by creating a critical defect (~5 mm diameter) in the central area of the calvaria bone. The pockets were cleaned with surgical gauze, filled with commercially available Bio-Oss®, and different bioceramics sizes for four weeks. The pocket with no sample was considered as a negative control. The soft tissue was repositioned and sutured carefully after the implantation of the materials. After the four weeks of treatment, the animals were sacrificed with isoflurane, and each sample was recovered and treated with 4% formalin (paraformaldehyde in 1× sterile PBS). No leaching was observed from one implant site to the neighboring site after four weeks of treatment. For histological analysis, hematoxylin and eosin (H&E) staining was performed. Decalcification of the newly developed calcified tissue was done with a 10% aqueous formic acid solution. The decalcified section images were taken by a light microscope (BX-50, Olympus Optical Co., Tokyo, Japan). All experiments were performed under the guideline of the approved animal protocol.

2.7. Micro-Computed Tomography (μCT) Analysis

The μCT was performed to analyze the prepared bioceramics in vivo bone regeneration efficiency, as previously described [26]. Briefly, the treated skull was excised, cleaned, and fixed by 4% PFA, and visualized by a μCT analyzer (Spectra Lago X, Tucson, AZ, USA) at 20 μm resolution. The images were captured, and the new bone formation was examined.

2.8. Statistical Analysis

Statistical analysis was carried with one-way ANOVA using the Origin Pro 9.0 software (Origin Pro v9.0, Origin Lab Corp., Northampton, MA, USA). All experiments were performed in triplicate ($n = 3$), and the results are expressed as mean ± standard deviation (SD). Statistical significance was considered as * $p < 0.05$.

3. Results

The microstructural characteristics of human teeth-derived bioceramics were examined through FE-SEM measurement, and the morphologies are presented in Figure 1a. The particle diameter of the prepared bioceramics occurred in the range of 50–500 nm (Figure 1b). The FE-SEM morphologies of SBF-soaked bioceramics after different time intervals are presented in Figure 1c,d. The SBF-soaked bioceramics exhibited the deposition of the granular particles on their surface. The elemental composition of commercially available HAp and prepared bioceramics was determined by X-ray fluorescence (XRF) spectroscopy, and the results are given in Table 1. It was noted that the Ca/P molar ratio was 1.67. Teeth powders contain some low-level metal impurities (i.e., Na, Mg, Al, K, Zn, Fe, Cu), which may be attributed to traces supplied by the chemical reagents or biomass trace elements. The EDX spectra of fresh and SBF-soaked bioceramics are shown in Figure 2a,b. The EDX result indicated that the bioceramics were mainly composed of calcium (Ca), phosphorus (Pa), and oxygen (O). The trace amounts of other elements, such as magnesium and chlorine, were also noted in the SBF-soaked bioceramics. The elemental analysis of fresh and 12-month SBF-soaked bioceramics through energy-dispersive X-ray (EDX) spectroscopy is given in Table 2. The EDX elemental mapping of human teeth-derived bioceramics is given in Figure 3. The obtained results indicated that the O, P, and Ca were the main constituents of the developed bioceramics, and they were uniformly distributed in the sample.

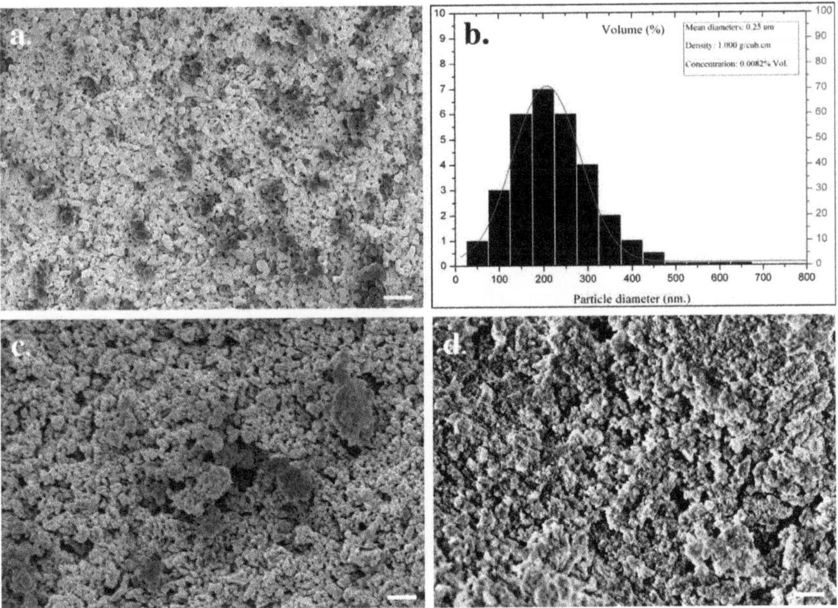

Figure 1. Surface morphology of human teeth-derived bioceramics through heat treatment; (**a**) the FE-SEM image of fresh sample; (**b**) particle size distribution curve of fresh bioceramics ($n = 3$, 50 particles); (**c**) the FE-SEM image of SBF-soaked bioceramics (6 months); and (**d**) the FE-SEM image of SBF-soaked bioceramics (12 months) (Scale bar = 10 μm).

Table 1. The determination of percentage element composition for commercially available HAp and human teeth-derived bioceramics through X-ray fluorescence (XRF) spectroscopy and their Ca/P molar ratio.

Product	$Ca_5(PO_4)_3(OH)$	Bioceramics
Ca	39.9	36.66
P	18.5	16.40
Ca/P Ratio	1.61	1.67
Na	-	0.79
Mg	-	0.60
Al	-	0.04
Si	-	0.07
Cu	-	0.01
Cl	-	0.20
K	-	0.03
Zn	-	0.14
Sr	-	0.03
Y	-	0.04
Zr	-	0.60
Ag	-	0.02
Fe	-	0.01
S	-	0.03

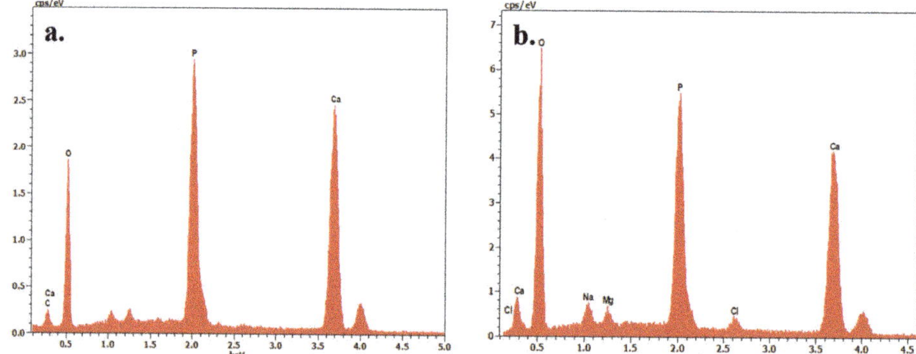

Figure 2. The energy-dispersive X-ray (EDX) profile of the prepared bioceramics; (**a**) fresh; and (**b**) SBF-soaked bioceramics (12 months).

Table 2. Elemental analysis of human-teeth derived bioceramics under different conditions through energy-dispersive X-ray (EDX) spectroscopy.

	Composition (%)	
Element	Fresh Bioceramics	SBF-Soaked Bioceramics (12 Months)
Calcium	35.70	40.50
Phosphorus	16.70	16.66
Oxygen	47.60	39.91
Sodium	-	1.23
Chlorine	-	1.09
Magnesium	-	0.61
Total	100.00	100.00

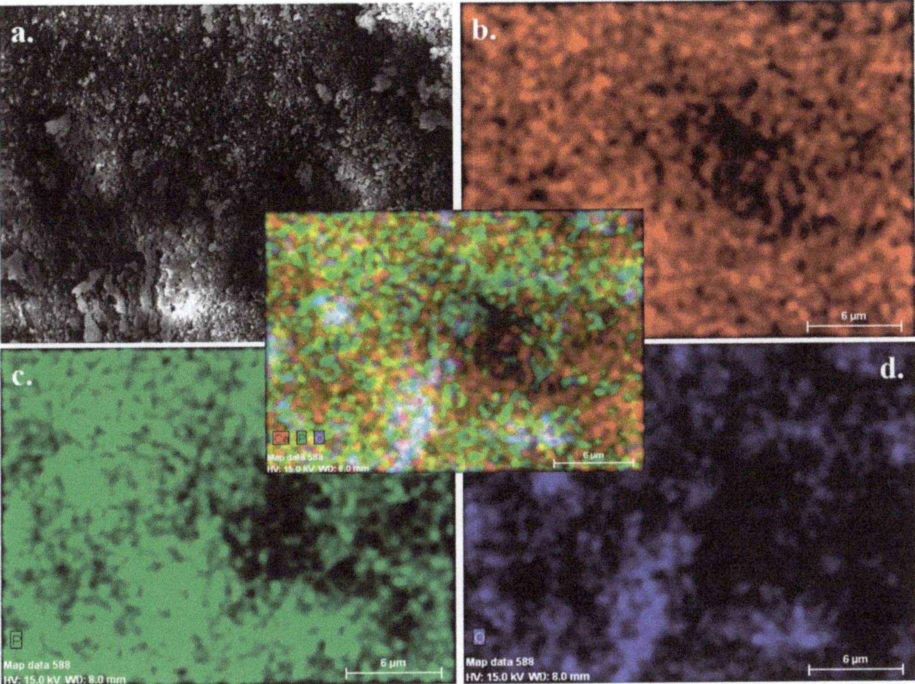

Figure 3. (**a**) The EDX elemental mapping of human teeth-derived bioceramics; (**b**) calcium; (**c**) phosphorous; and (**d**) oxygen (a merged image is shown at the center).

The XRD pattern of human teeth-derived bioceramics is shown in Figure 4. The XRD pattern resembled the previously reported pattern of HAp, indicating that heat treatment facilitates the formation of natural HAp from human teeth [27]. The XRD peak at 2θ = 33.3° suggested the existence of the CaP moiety in developed bioceramics [28]. The presence of the different functional groups in bioceramics was examined through the FTIR spectroscopy. The FTIR spectra of fresh and SBF-soaked (12 months) bioceramics are shown in Figure 5. The FTIR results are identical to the HAp pattern [27]. The appearance of several peaks in the regions of 3532, 2530–2329, 1640–1620, and 960–570 cm^{-1} matched the corresponding absorption peaks of HAp. The appearance of these peaks indicated the presence of different ions such as phosphate (PO_4^{3-}), hydroxyl (OH^-), and carbonate (CO_3^{2-}) in bioceramics. The thermal stability of the developed bioceramics was determined through TGA, and the obtained thermogram is given in Figure 6. The TGA curve shows the ~5% weight loss in the lower temperature region (up to 200 °C). A small endothermic transition was observed in this region in the DTA curve, but no noticeable weight loss was noted in the range of ~400 °C to 600 °C. However, a continuous weight loss was observed above 600 °C. The total weight loss of the bioceramics was ~12% at 1400 °C.

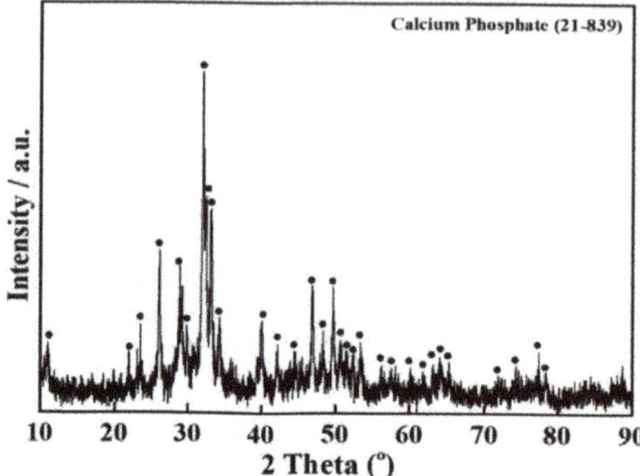

Figure 4. The XRD pattern of human teeth-derived bioceramics through heat treatment.

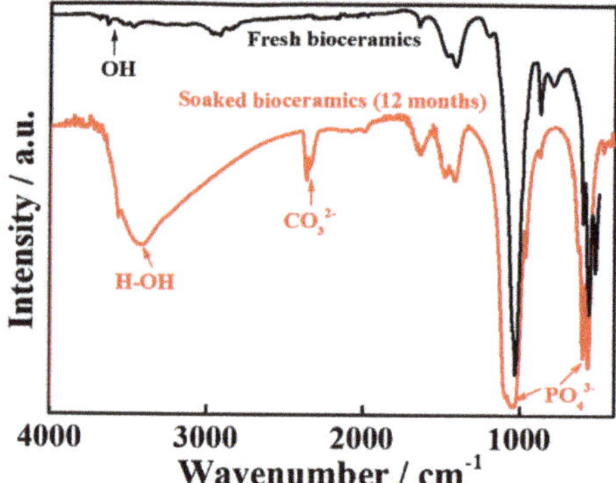

Figure 5. The FTIR spectra of fresh and SBF-soaked (12 months) bioceramics.

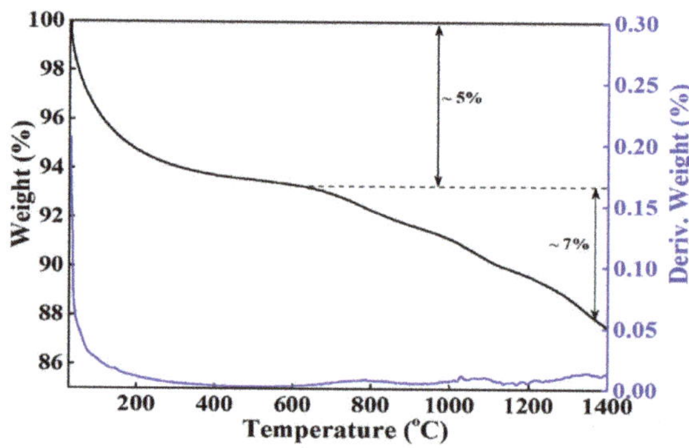

Figure 6. Thermal stability of human teeth-derived bioceramics; black line for TGA curve and blue line shows their derivative curve for fresh bioceramics under inert condition (N_2 atmosphere).

Cytotoxicity of fresh and SBF-soaked bioceramics (6 months) was monitored by WST-1 assay in the presence of hABMSCs after 24 h of incubation, and the results are given in Figure 7a. No adverse effects were observed on hABMSCs in the presence of fresh and SBF-soaked bioceramics, indicating their biocompatibility. The FE-SEM morphologies of the hABMSCs cultured on developed bioceramics at different periods are shown in Figure 7b,c. The cultured cells were healthy and proliferated under the experimental treatment, showing their biocompatibility. Bone regeneration potential of the developed bioceramics was evaluated through in vivo transplantation experiment using six-week-old imprinting control region (ICR) male mice. The groups without any treatment and with commercially available Bio-Oss® were considered as negative and positive control. The bioceramics (micro and nano-sized) treated groups were taken as experimental groups. The images of in vivo transplantation are shown in Figure 8a,b. No inflammation was observed around the transplanted area after four weeks of treatment, suggesting that the developed bioceramics were nontoxic and biocompatible. Rapid bone regeneration occurred in the presence of developed bioceramics, and this efficiency was intensely affected by the size of the developed bioceramics. The H&E staining process was performed to assess the histological analysis of the conducted experiment after four weeks of transplantation, and the results are given in Figure 9. The formation of new connective tissues and capillaries was observed around the transplanted area after four weeks of treatment, and its density was high in bioceramics-treated groups compared to the control and commercially available Bio-Oss® (Figure 9c,d). Various real-time non-invasive imaging techniques were applied to detect the bone-healing or correct placement during the implantation. These were X-ray computed tomography (CT), magnetic resonance imaging (MRI), and ultrasound (US), which can monitor the natural repair and the fate of host–materials interaction, in vivo [29]. CT images of the defected male mice skulls under different experimental conditions are shown in Figure 10. This work was different from our previous work, where we had evaluated the effects of sintering temperature on the formation of calcium-based bioceramics for tissue engineering [23].

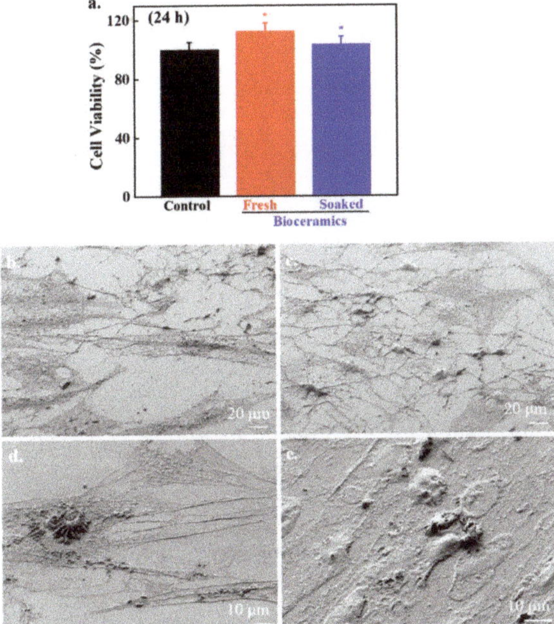

Figure 7. Cytotoxicity evaluation of human teeth-derived bioceramics; (**a**) indirect cell viability data after 24 h of treatment with the extracted media of fresh and SBF-soaked bioceramics (12 months); (**b,c**) the FE-SEM morphologies of the cultured cells in the presence of fresh bioceramics after 1 and 7 days of treatment at low magnification (1000×); and (**d,e**) high magnification (3000×), respectively.

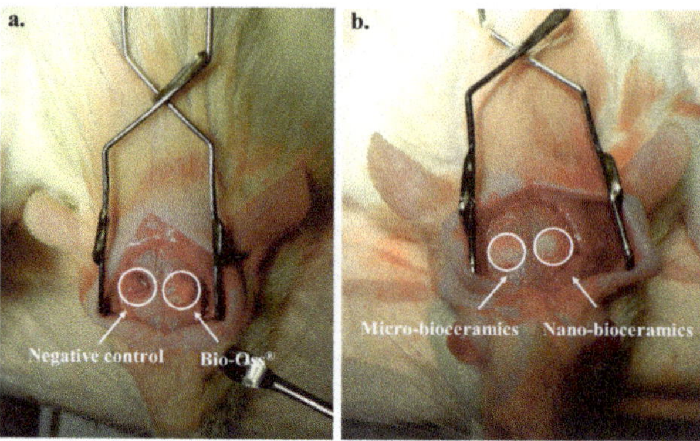

Figure 8. Images of in vivo transplantation; (**a**) with negative and positive control; and (**b**) in the presence of different sizes of bioceramics.

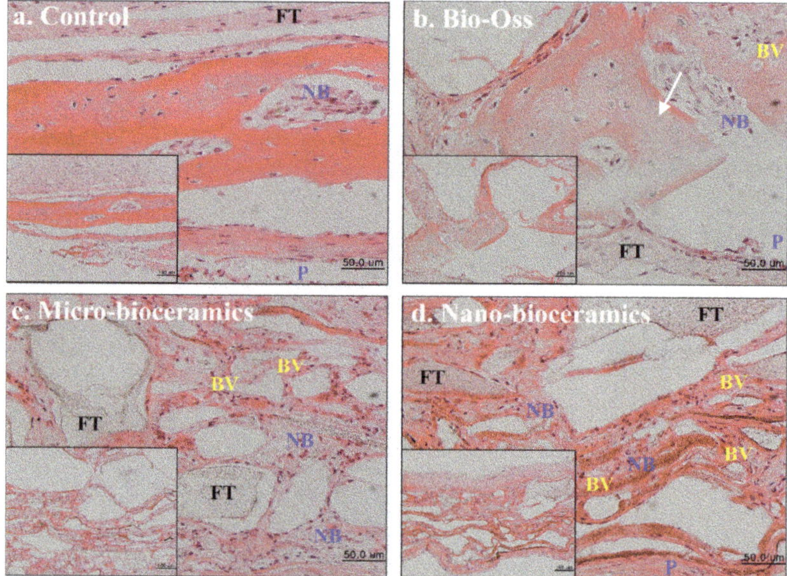

Figure 9. Hematoxylin and Eosin (H&E)-stained images of human-teeth derived bioceramics; (**a**) negative control; (**b**) positive control; (**c**) with micro and nano-sized bioceramics; and (**d**) in the presence of nano-sized bioceramics after four weeks of transplantation (Scale bar: 50 and 100 μm) Here, NB, P, BV, and FT indicate the new bone, periosteum, blood vessel, and fibrous tissue, respectively.

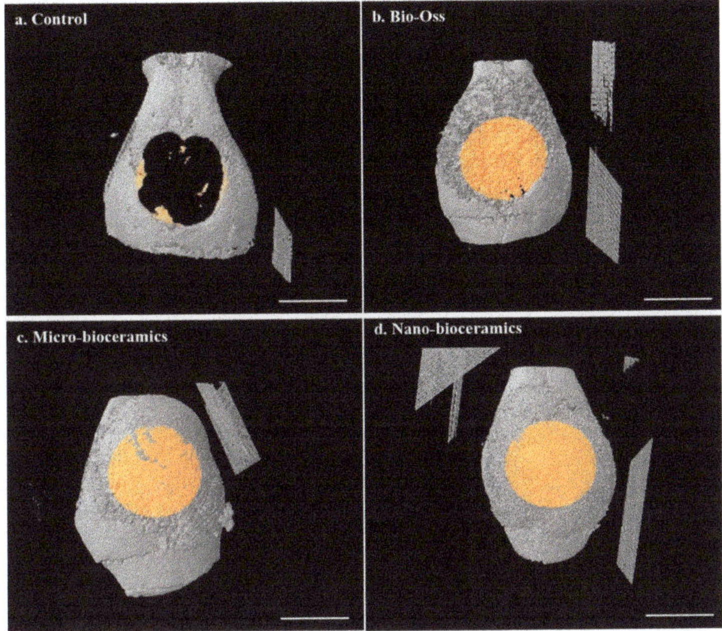

Figure 10. Micro-computed tomography (μCT) images of the defected mice skull in the presence of indicated materials after four weeks of transplantation; (**a**) control, (**b**) Bio-Oss, (**c**) micro-bioceramics, and (**d**) nano-bioceramics (Scale bar: 2 mm). The percentage bone healing was ~7 ± 2.08, 92 ± 3.27, 88 ± 4.70, and 95 ± 2.14% for the control, Bio-Oss, micro, and nano-bioceramics, respectively.

4. Discussion

Bone graft materials should have osteo-inductive potential to achieve optimal bone regeneration results. Their selection depends on clinical considerations and patient morbidity. The autogenous bone graft is considered the gold standard in the graft materials due to its inheritance osteo-inductive property. However, it has several disadvantages, including graft material availability, increased grafting time, blood loss, and additional cost [3]. The allogenic bone graft has better availability than autograft and does not require the second surgical procedure, but there is a chance of disease transmission and adverse immune responses, limiting its broad applicability [3]. Allogenic bone grafts are considered to be superiorly osteoconductive but weakly osteo-inductive and non-osteogenic [30]. Allogenic bone grafts can be obtained from living donors (femoral head), multiple organ donors, and post mortem donors (bone). The most important step in reducing immunogenicity and disease transmission is fluid pressurization to eliminate bone marrow and cellular debris. The detailed medical, social, sexual, and behavioral history of allograft donors is performed to further detect disease transmission risk factors. The number of the intraoral donor site depends on the quantity, geometry, and type of the bone required for reconstruction [31].

Therefore, we need an alternative and safe substitute for bone grafts. Synthetic bioceramics can be applied as a bone grafts substitute, but its limited availability and toxicity are serious concerns. Hence, finding an alternate source of biocompatible ceramics will increase their application in the tissue engineering sector. Herein, the bioceramics developed from human teeth were used as a bone reconstructing material in vivo. The fresh bioceramics demonstrated a particle-like morphology, while a rough surface and partially porous structure were in SBF-soaked bioceramics. The sintering kinetics played an important role in the size of the bioceramics. It was observed that the pore size and crystal morphology of material obtained from human and bovine bone samples were profoundly affected by the calcination temperature [32]. The bioactivity studies in SBF showed granular particle morphology due to the calcification of the samples in the media, and the ball-like apatite layer was also observed in the FE-SEM images, which proved the bioactivity of the samples. The elemental mapping analysis enabled the information related to the uniform distribution of the elements and the inhomogeneity in fresh and SBF-soaked bioceramics [33]. The appearance of the sharp peaks in the XRD pattern showed the crystalline nature of the developed bioceramics, which was extensively influenced by the sintering temperature. An enhancement in the crystallinity was reported earlier in bovine bone-derived HAp by increasing the calcination temperature 600 °C to 1000 °C [34]. A decrease in the crystallinity occurred above 1000 °C in Hap, owing to the decomposition of HAp into β-TCP [35,36]. Therefore, 1000 °C is assumed as a suitable calcination temperature for the formation of crystalline HAp. The Ca/P ratio in the developed bioceramics was determined by XRF analysis, and it was 1.67. This value is close to the commercially available HAp (1.61) [37]. SBF-soaked bioceramics exhibited a broad FTIR absorption peak of –OH compared to fresh bioceramics due to the presence of adsorbed moisture moiety [38]. As evidenced in the TGA curve, the weight loss was due to the removal of moisture moiety from the sample [39], further supported by a small endothermic transition in this region in the DTA curve. The continuous weight loss from the samples above 600 °C can be attributed to the dehydroxylation of HAp and its conversion to whitlockite [40]. No exothermic peaks occurred in the developed bioceramics, indicating the sample's high purity [41].

Fresh and SBF-soaked bioceramics had no adverse effects on hABMSCs, indicating their biocompatibility. It is well-known that cellular activity is significantly affected by material properties such as surface roughness, texture, and surface chemistry [42,43]. They can change the cell's proliferation by changing the physico-chemical interactions between the material surface and cells through ionic forces [44]. Cell adhesion potential offers the information associated with the biomaterial's possible use as an implant [43]. The FE-SEM images indicated that the cells were healthy and adhered adequately to flattened osteoblast-like morphology connected by the filopodia. The cell density was increased with increasing the culture time. In vivo study demonstrated that the bone regeneration efficiency was high in bioceramics-treated groups compared to the control, which was more significant in nano-sized bioceramics. This was probably due to their superior biocompatibility. The size effects of hydroxyapatite nanoparticles on human osteoblast-like MG-63 cells were also studied by Shi et al. They observed that the hydroxyapatite with the smallest diameter exhibited superior cellular activity in their macro-analogs [45]. It has been noted that nano-sized HAp incorporated composites mimicked natural bone conditions and supported better growth and proliferation of human osteoblast cells [46]. Nano-sized HAps are widely used in the field of tissue engineering for the rapid regeneration of bone tissues. However, selecting a suitable material for these applications still poses a challenging task for the researcher [47]. The osteogenic potential of calcium phosphate (CaP)-based bioceramics was earlier reported, and this potential is profoundly affected by the bone-like apatite layer formation and the absorption of protein on the surface of the materials [48,49]. The CaP-based bioceramics are considered bioactive materials and have excellent biocompatibility and osteo-conductivity potential. The phase composition and porosity extensively influence the osteo-conductivity of the materials.

HAp and other CaP-based bioceramics can regulate cell differentiation via the osteogenic process and promote bone tissue regeneration. Enhanced vascularization was observed in bioceramics-treated groups and their density further increased in nano-sized bioceramics-treated mice. Micro CT and histological analysis suggested that the defected mice skulls were nearly filled with new tissues in the presence of developed bioceramics compared to the control. Following the X-ray images, the micro and nano-sized bioceramics-treated groups showed better integration with the native bone after four weeks of transplantation (Figure 10). The HAp has a porous structure, facilitating the infiltration and migration of osteoblast cells from host bone to the defected sites [50–52]. No donor-to-donor variability issues were observed in the bioceramics developed to use as an implanted biomaterial. This study was limited to evaluating the bone regeneration potential of bioceramics derived from waste human teeth. However, more detailed clinical studies are required to assess any adverse effects of the developed bioceramics to allow their commercial applications. The limitations of biological-derived ceramic bone grafts are their batch variability, which can be minimized by the pooling of the donor sources [53]. It has been observed that the donor-specific variabilities such as cell proliferation, cell viability, and osteogenic differentiation occurred in human mesenchymal stromal cells (MSCs) obtained from different donors, which restricted their wide applicability. Several factors, like donor's age, sex, disease, and hormone status, are responsible for the donor-specific variability. Therefore, the pooling of human MSCs from different donors is applied to minimize the donor-specific variabilities. This approach has already been utilized in the past and can be used in bone tissue engineering [54]. Hence, our work demonstrates that heat treatment facilitates the formation of a useful bioactive material from waste human teeth for the rapid regeneration of bone tissues, and derived nano-sized bioceramics can be considered a suitable material for tissue engineering.

5. Conclusions

This study investigated the preparation and characterization of calcium phosphate-based bioceramics from human teeth. The FTIR spectra indicated the presence of characteristic peaks of the HAp, which was also confirmed through the XRD pattern. The appearance of sharp diffraction peaks in XRD suggested their crystallinity. The Ca/P molar ratio was determined through EDX and XRF techniques, and it was 1.60 and 1.67, respectively, which is closed to the commercially available hydroxyapatite (1.61). An enhancement in the Ca/P molar ratio (1.80) was observed in the SBF-soaked sample rather than in the fresh due to the calcification in SBF conditions. The particle size of the prepared bioceramics occurred in the range of approximately 50–500 nm diameter. No adverse effects were noted on hABMSCs in the presence of developed bioceramics, indicating their biocompatibility. The cells were healthy and appropriately adhered to the developed bioceramics.

Additionally, improved bone tissue regeneration occurred in the presence of prepared bioceramics rather than in the control, and commercially available Bio-Oss® demonstrated their superior osteogenic potential. This potential is significantly affected by the size of the bioceramics. Based on these findings, we concluded that nano-sized calcium phosphate-based bioceramics have great potential for bone tissue application.

Supplementary Materials: The following are available online at http://www.mdpi.com/2079-4991/10/12/2396/s1, Table S1: The chemical compositions of the SBF solution used in this study.

Author Contributions: K.-T.L., H.-W.C., and H.J. conceptualized the work and performed the synthesis, in vitro, and in-vivo experiments; J.H.C. monitored the experiment; S.D.D., A.B., and D.K.P. analyzed the data; D.K.P. wrote the original draft, and further modified and revised was by K.-T.L.; Funding acquisition by K.-T.L. and J.H.C. All authors have read and agreed to the published version of the manuscript.

Funding: This research was supported by the Basic Science Research Program through the National Research Foundation of Korea funded by the Ministry of Education (No. 2018R1A6A1A03025582 and 2019R1D1A3A03103828).

Conflicts of Interest: The authors declare no conflict of interest.

Ethical Statement: All experiments regarding animal tissue handling and culture were carried out as approved by the Institutional Review Board (IRB No. CRI05008) of the Dental Hospital, Seoul National University, Seoul, Korea.

References

1. Kurtz, S.; Ong, K.; Lau, E.; Mowad, F.; Halpern, M. Projections of Primary and Revision Hip and Knee Arthroplasty in the United States from 2005 to 2030. *J. Bone Joint Surg.* **2007**, *89*, 780–785. [CrossRef]
2. Bhattacharjee, A.; Gupta, A.; Verma, M.; Murugan, P.A.; Sengupta, P.; Matheshwaran, S.; Manna, I.; Balani, K. Site-Specific Antibacterial Efficacy and cyto/Hemo-Compatibility of Zinc Substituted Hydroxyapatite. *Ceram. Int.* **2019**, *45*, 12225–12233. [CrossRef]
3. Baylón, K.; Rodríguez-Camarillo, P.; Elías-Zúñiga, A.; Díaz-Elizondo, J.A.; Gilkerson, R.; Lozano, K. Past, Present and Future of Surgical Meshes: A Review. *Membranes* **2017**, *7*, 47. [CrossRef] [PubMed]
4. Kattimani, V.S.; Kondaka, S.; Lingamaneni, K.P. Hydroxyapatite—Past, present, and future in bone regeneration. *Bone Tissue Regen. Insights* **2016**, *7*, S36138. [CrossRef]

5. Khan, S.N.; Tomin, E.; Lane, J.M. Clinical applications of bone graft substitutes. *Orthop. Clin. N. Am.* **2000**, *31*, 389–398. [CrossRef]

6. Verbeeck, L.; Geris, L.; Tylzanowski, P.; Luyten, F.P. Uncoupling of in-Vitro Identity of Embryonic Limb Derived Skeletal Progenitors and Their in-vivo Bone Forming Potential. *Sci. Rep.* **2019**, *9*, 5782. [CrossRef]

7. Cheng, L.; Cai, Z.; Zhao, J.; Wang, F.; Lu, M.; Deng, L.; Cui, W. Black Phosphorus-Based 2D Materials for Bone Therapy. *Bioact. Mater.* **2020**, *5*, 1026–1043. [CrossRef]

8. Patel, D.K.; Jin, B.; Dutta, S.D.; Lim, K.-T. Osteogenic Potential of Human Mesenchymal Stem Cells on Eggshells-Derived Hydroxyapatite Nanoparticles for Tissue Engineering. *J. Biomed. Mater. Res. Part B Appl. Biomater.* **2019**, *108*, 1953–1960. [CrossRef]

9. Hornez, J.-C.; Chai, F.; Monchau, F.; Blanchemain, N.; Descamps, M.; Hildebrand, H. Biological and Physico-Chemical Assessment of Hydroxyapatite (HA) With Different Porosity. *Biomol. Eng.* **2007**, *24*, 505–509. [CrossRef]

10. Ferraz, M.P.; Monteiro, F.J.; Manuel, C.M. Hydroxyapatite Nanoparticles: A Review of Preparation Methodologies. *J. Appl. Biomater. Biomech.* **2010**, *2*, 74–80.

11. Bu, S.; Yan, S.; Wang, R.; Xia, P.; Zhang, K.; Li, G.; Yin, J. In Situ Precipitation of Cluster and Acicular Hydroxyapatite onto Porous Poly(γ-Benzyl-l-Glutamate) Microcarriers for Bone Tissue Engineering. *ACS Appl. Mater. Interfaces* **2020**, *12*, 12468–12477. [CrossRef] [PubMed]

12. Wu, H.; Xie, L.; He, M.; Zhang, R.; Tian, Y.; Liu, S.; Gong, T.; Huo, F.; Yang, T.; Zhang, Q.; et al. A Wear-Resistant TiO_2 Nanoceramic Coating on Titanium Implants for Visible-Light Photocatalytic Removal of Organic Residues. *Acta Biomater.* **2019**, *97*, 597–607. [CrossRef]

13. Londoño-Restrepo, S.M.; Jeronimo-Cruz, R.; Millán-Malo, B.M.; Rivera-Muñoz, E.M.; Rodriguez-García, M.E. Effect of the Nano Crystal Size on the X-Ray Diffraction Patterns of Biogenic Hydroxyapatite from Human, Bovine, and Porcine Bones. *Sci. Rep.* **2019**, *9*, 1–12. [CrossRef] [PubMed]

14. Beniash, E.; Stifler, C.A.; Sun, C.-Y.; Jung, G.S.; Qin, Z.; Buehler, M.J.; Gilbert, P.U.P.A. The Hidden Structure of Human Enamel. *Nat. Commun.* **2019**, *10*, 1–13. [CrossRef] [PubMed]

15. Nayak, B.; Samant, A.; Misra, P.K.; Saxena, M. Nanocrystalline Hydroxyapatite: A Potent Material for Adsorption, Biological and Catalytic Studies. *Mater. Today Proc.* **2019**, *9*, 689–698. [CrossRef]

16. Huang, C.; Bhagia, S.; Hao, N.; Meng, X.; Liang, L.; Yong, Q.; Ragauskas, A.J. Biomimetic Composite Scaffold from an in situ Hydroxyapatite Coating on Cellulose Nanocrystals. *RSC Adv.* **2019**, *9*, 5786–5793. [CrossRef]

17. Bayani, M.; Torabi, S.; Shahnaz, A.; Pourali, M. Main Properties of Nanocrystalline Hydroxyapatite as a Bone Graft Material in Treatment of Periodontal Defects. A Review of Literature. *Biotechnol. Biotechnol. Equip.* **2017**, *31*, 215–220. [CrossRef]

18. Wohlhauser, S.; Delepierre, G.; Labet, M.; Morandi, G.; Thielemans, W.; Weder, C.; Zoppe, J.O. Grafting PolymersfromCellulose Nanocrystals: Synthesis, Properties, and Applications. *Macromolecules* **2018**, *51*, 6157–6189. [CrossRef]

19. Pillai, S.C.; Lang, Y. *Toxicity of Nanomaterials: Environmental and Healthcare Applications*; CRC Press: Boca Raton, FL, USA, 2019.

20. Gogoi, S.; Das, B.; Chattopadhyay, D.; Khan, R. Sustainable Nanostructural Materials for Tissue Engineering. In *Dynamics of Advanced Sustainable Nanomaterials and Their Related Nanocomposites at the Bio-Nano Interface*; Elsevier: Amsterdam, The Netherlands, 2019; pp. 75–100.

21. Ruksudjarit, A.; Pengpat, K.; Rujijanagul, G.; Tunkasiri, T. Synthesis and Characterization of Nanocrystalline Hydroxyapatite from Natural Bovine Bone. *Curr. Appl. Phys.* **2008**, *8*, 270–272. [CrossRef]

22. Pu'Ad, N.M.; Koshy, P.; Abdullah, H.; Idris, M.; Lee, T. Syntheses of Hydroxyapatite from Natural Sources. *Heliyon* **2019**, *5*, e01588.

23. Lim, K.-T.; Suh, J.D.; Kim, J.; Choung, P.-H.; Chung, J.H. Calcium Phosphate Bioceramics Fabricated from Extracted Human Teeth for Tooth Tissue Engineering. *J. Biomed. Mater. Res. Part B Appl. Biomater.* **2011**, *99*, 399–411. [CrossRef] [PubMed]

24. Gronthos, S.; Mankani, M.; Brahim, J.; Robey Gehron, P.; Shi, S. Postnatal Human Dental Pulp Stem Cells (DPSCs) in Vitro and in Vivo. *Proc. Natl. Acad. Sci. USA* **2000**, *97*, 13625–13630. [CrossRef]

25. Sangsanoh, P.; Waleetorncheepsawat, S.; Suwantong, O.; Wutticharoenmongkol, P.; Weeranantanapan, O.; Chuenjitbuntaworn, B.; Cheepsunthorn, P.; Pavasant, A.P.; Supaphol, P. In Vitro Biocompatibility of Schwann Cells on Surfaces of Biocompatible Polymeric Electrospun Fibrous and Solution-Cast Film Scaffolds. *Biomacromolecules* **2007**, *8*, 1587–1594. [CrossRef] [PubMed]

26. Patel, D.K.; Dutta, S.D.; Hexiu, J.; Ganguly, K.; Lim, K.-T. Bioactive Electrospun Nanocomposite Scaffolds of poly(lactic acid)/Cellulose Nanocrystals for Bone Tissue Engineering. *Int. J. Biol. Macromol.* **2020**, *162*, 1429–1441. [CrossRef] [PubMed]

27. Elkayar, A.; El Shazly, Y.; Assaad, M. Properties of Hydroxyapatite from Bovine Teeth. *Bone Tissue Regen. Insights* **2009**, *2*, S3728. [CrossRef]

28. Lim, K.-T.; Patel, D.K.; Choung, H.W.; Seonwoo, H.; Kim, J.; Chung, J.H. Evaluation of Bone Regeneration Potential of Long-Term Soaked Natural Hydroxyapatite. *ACS Appl. Bio Mater.* **2019**, *2*, 5535–5543. [CrossRef]

29. Fragogeorgi, E.; Rouchota, M.; Georgiou, M.; Velez, M.; Bouziotis, P.; Loudos, G. In Vivo Imaging Techniques for Bone Tissue Engineering. *J. Tissue Eng.* **2019**, *10*, 2041731419854586. [CrossRef] [PubMed]

30. Suchomel, P.; Barsa, P.; Buchvald, P.; Svobodnik, A.; Vanickova, E. Autologous Versus Allogenic Bone Grafts in Instrumented Anterior Cervical Discectomy and Fusion: A Prospective Study with Respect to Bone Union Pattern. *Eur. Spine J.* **2004**, *13*, 510–515. [CrossRef]

31. Sakkas, A.; Wilde, F.; Heufelder, M.; Winter, K.; Schramm, A. Autogenous Bone Grafts in Oral implantology—Is It Still a "Gold standard"? A Consecutive Review of 279 Patients with 456 Clinical Procedures. *Int. J. Implant. Dent.* **2017**, *3*, 1–17. [CrossRef]

32. Figueiredo, M.J.D.F.M.D.; Fernando, A.; Martins, G.B.; Freitas, J.C.C.; Judas, F. Effect of the Calcination Temperature on the Composition and Microstructure of Hydroxyapatite Derived from Human and Animal Bone. *Ceram. Int.* **2010**, *36*, 2383–2393. [CrossRef]

33. Predoi, D.; Iconaru, S.L.; Predoi, D.; Motelica-Heino, M.; Guégan, R.; Buton, N. Evaluation of Antibacterial Activity of Zinc-Doped Hydroxyapatite Colloids and Dispersion Stability Using Ultrasounds. *Nanomaterials* **2019**, *9*, 515. [CrossRef] [PubMed]

34. Niakan, A.; Ramesh, S.; Ganesan, P.; Tan, C.; Purbolaksono, J.; Chandran, H.; Teng, W. Sintering Behaviour of Natural Porous Hydroxyapatite Derived from Bovine Bone. *Ceram. Int.* **2015**, *41*, 3024–3029. [CrossRef]

35. Jinlong, N.; Zhenxi, Z.; Dazong, J. Investigation of Phase Evolution during the Thermochemical Synthesis of Tricalcium Phosphate. *J. Mater. Synth. Process.* **2001**, *9*, 235–240. [CrossRef]

36. De Oliviera Ugarte, J.F.; De Sena, L.A.; de Castro Pérez, C.A.; De Aguiar, P.F.; Rossi, A.M.; Soares, G.A. Influence of Processing Parameters on Structural Characteristics of Porous Calcium Phosphate Samples: A Study Using an Experimental Design Method. *Mater. Res.* **2005**, *8*, 71–76. [CrossRef]

37. Handley-Sidhu, S.; Renshaw, J.C.; Yong, P.; Kerley, R.; Macaskie, L. Nano-Crystalline Hydroxyapatite Bio-Mineral for the Treatment of Strontium from Aqueous Solutions. *Biotechnol. Lett.* **2010**, *33*, 79–87. [CrossRef] [PubMed]

38. Lim, K.-T.; Baik, S.J.; Kim, S.W.; Kim, J.; Seonwoo, H.; Kim, J.-W.; Choung, P.-H.; Choung, Y.-H.; Chung, J.H. Development and Characterization of Fast-Hardening Composite Cements Composed of Natural Ceramics Originated from Horse Bones and Chitosan Solution. *Tissue Eng. Regen. Med.* **2014**, *11*, 362–371. [CrossRef]

39. Ikeda, T.; Kasai, M.; Tatsukawa, E.; Kamitakahara, M.; Shibata, Y.; Yokoi, T.; Nemoto, T.K.; Ioku, K. A Bone Substitute with High Affinity for Vitamin D-binding Protein—Relationship with Niche of Osteoclasts. *J. Cell. Mol. Med.* **2014**, *18*, 170–180. [CrossRef]

40. Meejoo, S.; Maneeprakorn, W.; Winotai, P. Phase and Thermal Stability of Nanocrystalline Hydroxyapatite Prepared via Microwave Heating. *Thermochim. Acta* **2006**, *447*, 115–120. [CrossRef]

41. Shavandi, A.; Bekhit, A.E.-D.A.; Ali, A.; Sun, Z. Synthesis of Nano-Hydroxyapatite (nHA) from Waste Mussel Shells Using a Rapid Microwave Method. *Mater. Chem. Phys.* **2015**, *149*, 607–616. [CrossRef]

42. Patel, D.K.; Rana, D.; Aswal, V.K.; Srivastava, S.; Roy, P.; Maiti, P. Influence of Graphene on Self-Assembly of Polyurethane and Evaluation of Its Biomedical Properties. *Polymer* **2015**, *65*, 183–192. [CrossRef]

43. Dutta, S.D.; Patel, D.K.; Seo, Y.-R.; Park, C.-W.; Lee, S.-H.; Kim, J.-W.; Kim, J.; Seonwoo, H.; Lim, K.-T. In Vitro Biocompatibility of Electrospun Poly(ε-Caprolactone)/Cellulose Nanocrystals-Nanofibers for Tissue Engineering. *J. Nanomater.* **2019**, *2019*, 1–11. [CrossRef]

44. Leenaerts, O.; Partoens, B.; Peeters, F.M. Water on Graphene: Hydrophobicity and Dipole Moment Using Density Functional Theory. *Phys. Rev. B* **2009**, *79*, 235440. [CrossRef]

45. Shi, Z.; Huang, X.; Cai, Y.; Tang, R.; Yang, D. Size Effect of Hydroxyapatite Nanoparticles on Proliferation and Apoptosis of Osteoblast-Like Cells. *Acta Biomater.* **2009**, *5*, 338–345. [CrossRef] [PubMed]

46. Huang, J.; Lin, Y.W.; Fu, X.W.; Best, S.M.; Brooks, R.A.; Rushton, N.; Bonfield, W. Development of Nano-Sized Hydroxyapatite Reinforced Composites for Tissue Engineering Scaffolds. *J. Mater. Sci. Mater. Electron.* **2007**, *18*, 2151–2157. [CrossRef] [PubMed]

47. Wei, X.; Lee, Y.-K.; Huh, K.M.; Kim, S.; Park, K. *Safety and Efficacy of Nano/Micro Materials*; Springer: Berlin/Heidelberg, Germany, 2008; pp. 63–88.

48. Daculsi, G.; LeGeros, R.Z.; Deudon, C. Scanning and Transmission Electron Microscopy, and Electron Probe Analysis of the Interface Between Implants and Host Bone. Osseo-Coalescence Versus Osseo-Integration. *Scanning Microsc.* **1990**, *4*, 309–314.

49. Okumura, M.; Ohgushi, H.; Dohi, Y.; Katuda, T.; Tamai, S.; Koerten, H.K.; Tabata, S. Osteoblastic Phenotype Expression on the Surface of Hydroxyapatite Ceramics. *J. Biomed. Mater. Res.* **1997**, *37*, 122–129. [CrossRef]

50. Zhu, Y.; Zhang, K.; Zhao, R.; Ye, X.; Chen, X.; Xiao, Z.; Yang, X.; Zhu, X.; Zhang, K.; Fan, Y.; et al. Bone Regeneration with micro/Nano Hybrid-Structured Biphasic Calcium Phosphate Bioceramics at Segmental Bone Defect and the Induced Immunoregulation of MSCs. *Biomaterials* **2017**, *147*, 133–144. [CrossRef]

51. He, Y.; Liu, A.; Ke, X.; Sun, M.; He, Y.; Yang, X.; Fu, J.; Zhang, L.; Yang, G.; Liu, Y.; et al. 3D Robocasting Magnesium-Doped wollastonite/TCP Bioceramic Scaffolds with Improved Bone Regeneration Capacity in Critical Sized Calvarial Defects. *J. Mater. Chem. B* **2017**, *5*, 2941–2951.

52. Mansour, A.; Abu-Nada, L.; Al-Waeli, H.; Mezour, M.A.; Abdallah, M.; Kinsella, J.M.; Kort-Mascort, J.; Henderson, J.E.; Luna, J.L.R.-G.; Tran, S.D.; et al. Bone Extracts Immunomodulate and Enhance the Regenerative Performance of Dicalcium Phosphates Bioceramics. *Acta Biomater.* **2019**, *89*, 343–358. [CrossRef]

53. Diaz-Rodriguez, P.; Lopezalvarez, M.R.; Serra, J.; González, P.; Landín, M. Current Stage of Marine Ceramic Grafts for 3D Bone Tissue Regeneration. *Mar. Drugs* **2019**, *17*, 471. [CrossRef]

54. Widholz, B.; Tsitlakidis, S.; Reible, B.; Moghaddam, A.; Westhauser, F. Pooling of Patient-Derived Mesenchymal Stromal Cells Reduces Inter-Individual Confounder-Associated Variation Without Negative Impact on Cell Viability, Proliferation and Osteogenic Differentiation. *Cells* **2019**, *8*, 633. [CrossRef] [PubMed]

Publisher's Note: MDPI stays neutral with regard to jurisdictional claims in published maps and institutional affiliations.

 *nanomaterials*

Article

The Mechanosensing and Global DNA Methylation of Human Osteoblasts on MEW Fibers

Pingping Han [1,2], Cedryck Vaquette [2], Abdalla Abdal-hay [2,3] and Sašo Ivanovski [1,2,*]

1 Center for Oral-Facial Regeneration, Rehabilitation and Reconstruction (COR3), Epigenetics Nanodiagnostic and Therapeutic Group, School of Dentistry, The University of Queensland, Brisbane, QLD 4006, Australia; p.han@uq.edu.au
2 School of Dentistry, The University of Queensland, Herston, QLD 4006, Australia; c.vaquette@uq.edu.au (C.V.); abdalla.ali@uq.edu.au (A.A.-h.)
3 Department of Mechanical Engineering, Faculty of Engineering, South Valley University, Qena 83523, Egypt
* Correspondence: s.ivanovski@uq.edu.au

Abstract: Cells interact with 3D fibrous platform topography via a nano-scaled focal adhesion complex, and more research is required on how osteoblasts sense and respond to random and aligned fibers through nano-sized focal adhesions and their downstream events. The present study assessed human primary osteoblast cells' sensing and response to random and aligned medical-grade polycaprolactone (PCL) fibrous 3D scaffolds fabricated via the melt electrowriting (MEW) technique. Cells cultured on a tissue culture plate (TCP) were used as 2D controls. Compared to 2D TCP, 3D MEW fibrous substrates led to immature vinculin focal adhesion formation and significantly reduced nuclear localization of the mechanosensor-yes-associated protein (YAP). Notably, aligned MEW fibers induced elongated cell and nucleus shape and highly activated global DNA methylation of 5-methylcytosine, 5-hydroxymethylcytosine, and N-6 methylated deoxyadenosine compared to the random fibers. Furthermore, although osteogenic markers (*osterix-OSX* and *bone sialoprotein-BSP*) were significantly enhanced in PCL-R and PCL-A groups at seven days post-osteogenic differentiation, calcium deposits on all seeded samples did not show a difference after normalizing for DNA content after three weeks of osteogenic induction. Overall, our study linked 3D extracellular fiber alignment to nano-focal adhesion complex, nuclear mechanosensing, DNA epigenetics at an early point (24 h), and longer-term changes in osteoblast osteogenic differentiation.

Keywords: microgeometry; mechanobiology; global DNA methylation; osteoblast mechanosensing

Citation: Han, P.; Vaquette, C.; Abdal-hay, A.; Ivanovski, S. The Mechanosensing and Global DNA Methylation of Human Osteoblasts on MEW Fibers. *Nanomaterials* **2021**, *11*, 2943. https://doi.org/10.3390/nano11112943

Academic Editor: Antonio Gloria

Received: 24 September 2021
Accepted: 26 October 2021
Published: 3 November 2021

Publisher's Note: MDPI stays neutral with regard to jurisdictional claims in published maps and institutional affiliations.

1. Introduction

In the native bone milieu, osteoblasts can sense their local structural microenvironment to form focal adhesions and initiate "outside-in" mechanotransductive signals, leading to extracellular matrix formation along the cellular direction [1–3]. Biomaterial substrates with two dimensional (2D) or three dimensional (3D) structures as biophysical cues have a critical influence on mammalian cell phenotype, proliferation, adhesion, and differentiation [1,4–7]. Comprehensive studies have demonstrated that 2D substrates generate "outside-in" mechanotransductive signals that permit cells to sense and respond dynamically to their surrounding microenvironment [8,9]. Specifically, a 2D nanoscale grooved model established on titanium-alloy substrates by laser ablation demonstrated that the unique bone matrix formation orthogonal to mouse osteoblast cell alignment was facilitated by 3D nano-sized focal adhesion (FAs) maturation [2,3]. Furthermore, a previous study showed that 2D collagen hydrogels regulate histone modification and non-coding RNA [1]. Forces that are generated by nano-scaled FAs (such as vinculin [10]) are then transmitted to the cytoplasm, nucleus (via nuclear mechanosensor–YAP/TAZ [11]) and chromatin, through physical links on the nuclear membrane and the mesh-structured nuclear Lamin A/C network [12]. This process, in turn, may govern epigenetic mechanisms,

89

as well as transcription activation and repression, thus influencing cell behavior and cell differentiation fate [1,13,14].

Epigenetics (including DNA methylation, histone modification, and non-coding RNA) refers to heritable changes in gene expression that do not involve changes to the DNA sequence, but rather act through chemical alterations of DNA that induce rapid inactivation or activation of genes [15,16]. DNA methylation involves the addition of methyl groups to either cytosine or adenine at specific sites in the DNA sequence and is dynamically regulated by methyl writing enzymes (DNA methyltransferases (DNMTs)) and methyl erasing enzymes (ten-eleven translocation methylcytosine dioxygenase (TETs)) [16,17]. The most common form of DNA methylation is 5 methylcytosine (5mC) and 5mC can be demethylated by TETs to become 5-hydroxymethylcytosine (5hmC). Beyond 5mC and 5hmC, N6-methyladenosine modification in DNA (m6dA) is the most common DNA modification. The evidence shows that DNA methylation is associated with osteogenic differentiation of mesenchymal stem cells [18,19] and periodontitis disease pathogenesis. [20] The concept of 2D biomaterial physical features acting as epigenetic cues has been proposed [13,21], thus, it is important to investigate the effect of a micro-scaled fiber arrangement (random or aligned) within the 3D structure of the biomaterials that can lead to DNA hypo-or hyper-methylation when compared to a 2D substrate.

The biomaterial-based 3D structure provides a closer representation of in vivo conditions compared with a conventional 2D system. A 3D poly(l-lactide-co-caprolactone) (PLCL) membrane with aligned nanofibers induced an increased histone H3 acetylation and H3 methylation compared to random fibers in mouse ear fibroblasts [22]. However, the sensing and response of human osteoblasts to micro-scaled fiber deposition (aligned and random) remain poorly understood. The present study aimed to investigate 3D mechanobiology in terms of focal adhesion and nuclear mechanotransduction, as well as global DNA epigenetics in human osteoblasts. In this study, a medical-grade polycaprolactone (PCL) fibrous scaffold with random and aligned fibers as 3D substrates was fabricated by melt electrowriting (MEW) technology. MEW precisely places continuous PCL fibers onto a collector plate, as dictated by computer-aided design software [23,24]. PCL was selected due to its relative abundance, low cost, low immunogenicity reactions [25–27], FDA approval for use in humans [24], and more recently, for its use in tissue regeneration applications. Our study not only provides new insights in understanding the roles of cellular mechanics in modulating the cell fate and cell differentiation of osteoblasts, but also paves the way for using PCL 3D MEW fibrous substrate cell cultures as an ideal platform to investigate 3D mechanobiology and epigenetics in osteoblasts.

2. Materials and Methods

2.1. 3D PCL Porous Membranes Fabrication and Characterization

The PCL 3D fibrous substrates with two geometries (random and aligned fibers) were manufactured using medical-grade polycaprolactone (PCL, Purasorb PC 12), as previously described [28]. Briefly, melt electrowriting mode was utilized using a melt electrospinner: spinneret size of 23 G, temperature of 75 °C, an extrusion pressure of 180 kPa, a distance spinneret-collector of 10 mm, a voltage of 8 kV, and a stage speed of 1000 mm/min. The random fiber (PCL-R) fibrous substrates were fabricated by reducing the stage speed to 200 mm/min to allow buckling of the jet to obtain a random pattern. Aligned fibers (PCL-A) were fabricated by collector modifications where the theoretical pore size was 200 μm. The thickness of both PCL-A and PCL-R was approximately 80 μm.

Scanning electron microscopy (SEM; JEOL 7001f, NeoScope JCM-5000 Benchtop SEM, JEOL Pty. Ltd., Sydney, Australia) operating at a voltage of 5 kV was used to characterize the fabricated 3D fibrous substrate morphology following gold coating for 75 s (~10 nm coating thickness). The fiber diameter and pore size were calculated using ImageJ software (1.51V, National Institutes of Health, USA) as described previously [28]. Briefly, SEM images and optical images were taken and fiber center-to-center distance was measured as the pore size, along with fiber diameter.

For mechanical testing, the sample (6 mm long and 10 mm wide) was placed into the grips and tested using a Univert universal tester (Cellscale, Waterloo, Canada) fitted with a 10 N load cell and at a cross-head speed of 5 mm/min at room temperature. Three replicate samples were tested for each design. The tensile modulus was calculated from the initial linear section of the tensile curve from stain ranging from 1–8%).

2.2. Human Primary Osteoblast Culture and Differentiation

The isolation and culture of primary human alveolar bone-derived osteoblasts (hOBs) were performed according to published protocols [29–31]. All participants involved who underwent third molar extraction gave informed consent and the Human Ethics Committees of the University of Queensland approved the research protocol (Approval number 2019000134). hOBs were cultured in Dulbecco's modified Eagle's medium (DMEM; Gibco-Invitrogen) supplemented with 10% vol/vol fetal bovine serum (FBS; Thermo Scientific Australia, Sydney, Australia) and 50 U/mL penicillin and 50 mg/mL streptomycin (P/S; Gibco-Invitrogen) at 37 °C in a humidified CO_2 incubator. For osteogenic differentiation, cells were cultured in an osteogenic DMEM medium containing 10% FBS, 50 µg/mL ascorbic acid, 3 mM β-glycerophosphate, and 10 nM dexamethasone (Sigma-Aldrich, St. Louis, MO, USA). Cells sourced from two patients at passages 4–5 were used for all experiments. Cells cultured on polystyrene tissue culture plates (TCP) were used as 2D controls.

3D PCL fibrous substrates were treated with 2 N sodium hydroxide (NaOH) for 1 h before coating with FBS in order to enhance cell seeding efficacy, as described previously [32].

For immunostaining and global DNA epigenetics, the cells were seeded at a density of 3000 cells per 1×1 cm^2 porous membrane for 24 h.

2.3. Immunofluorescence Staining

Immunofluorescent staining was performed as previously described [8,33,34]. Briefly, hOBs were fixed in 4% PFA at room temperature for 10 min and then permeabilized with 0.05% Triton X-100 in PBS containing 320 mM sucrose and 6 mM magnesium chloride. After blocking with 1% bovine serum albumin, 0.1% Tween-20, 0.3 M glycine, 10% goat serum (Gibco) in PBS for 1 h, primary antibodies (Vinculin, 1:400, sc-73614, Santa Cruz Biotechnology; Lamin A/C, 1:200, sc-376248, Santa Cruz Biotechnology; YAP, sc-101199, Santa Cruz Biotechnology, Dallas, TX, USA) were incubated for 1 h at room temperature. After three PBS-Tween 20 washes, AlexaFluor-488 [H+L] secondary antibodies supplemented with AlexaFluor-conjugated phalloidin and DAPI were added for 1 h at room temperature. Images (>15 cells) were acquired with a Nikon confocal microscope (Nikon, Tokyo, Japan).

2.4. Focal Adhesion, Nucleus and Nuclear YAP Data Analysis

For morphological analysis of focal adhesion, vinculin staining images (>15 cells) were processed and analyzed in Fiji-ImageJ software (1.51V, National Institutes of Health, USA, v1.8.0). In this analysis, FAs were detected as bright clusters of vinculin, with FA numbers, percentage of FAs containing F-actin and FAs length [8]. Nucleus area, circularity, and aspect ratio (major axis/minor axis) were calculated from LAMIN A/C images using ImageJ software. Nuclear YAP/TAZ (%) were scored as predominantly nuclear versus evenly distributed/predominantly cytoplasmic, as previously described [8].

2.5. DNA Isolation and Global DNA Methylation Analysis

Genomic DNA was extracted using the PureLink™ Genomic DNA Mini Kit (Invitrogen™, ThermoFisher Scientific, Sydney, Australia) according to the manufacturer's instructions. Briefly, the cells from the various groups (three replicates) were suspended in lysis buffer and RNase A and Proteinase K were added to remove the protein and RNA. The quality and quantity of the DNA were determined by a NanoDrop One spectrophotometer (ThermoFisher Scientific, Waltham, MA, USA).

Global methylation analysis of 5 methyl cytosine (5mC), 5-hydroxymethylcytosine (5hmC), and N 6-methyladenosine (m6A) for DNA was performed by using a Global DNA Methylation Assay Kit (5mC, ab233486, Abcam, Cambridge, UK) Global DNA Hydroxymethylation Assay Kit (5hmc, ab233487, Abcam, Cambridge, UK), and an m6A DNA Methylation Assay Kit (ab233488, Abcam, Cambridge, UK), as per the manufacturer's instructions. Briefly, sample DNA positive controls at six different concentrations (to generate a standard curve) and negative control were mixed with DNA binding solution and incubated at 37 °C for 60 min. After washing three times with 150 μL washing buffer, 5mC/5hmC/m6 antibodies, along with a signal indicator and enhancer solution, were added and incubated at room temperature for 1 h. After washing with wash buffer five times, 50 μL developer solution was added and incubated for 3 min at room temperature until the positive control with the highest concentration turned blue. Subsequently, 50 μL of stop solution was added to each well for 2 min to stop the enzyme reaction. The absorbance was measured at 450 nm within 2 min on an Infinite Pro spectrometer.

The global methylation level of all DNAs was calculated using the following equations as described previously [20]:

$$5mc/5hmC\ \% = \frac{\text{Sample OD} - \text{Negative Control OD}}{\text{Slope} \times \text{S}} \times 100\% \tag{1}$$

$$m6A\ \% = \frac{(\text{Sample OD} - \text{Negative Control OD})/\text{S}}{(\text{Positive Control OD} - \text{NDC OD})/\text{P}} \times 100\% \tag{2}$$

where, the slope (OD/1%) was determined from the standard curve using linear regression; S is the amount of input sample DNA in ng; P is the amount of positive input control in ng; OD is the optical density.

2.6. RNA Isolation and Quantitative Real-Time Polymerase Chain Reaction (qPCR)

Quantitative reverse transcription-polymerase chain reaction (RT-qPCR) was used to measure the mRNA expression of osteogenic markers: *alkaline phosphatase (ALP), osteopontin (OPN), runt-related transcription factor 2 (RUNX2), OSX,* and *BSP.* The primers are listed in Table 1. Total RNA was isolated using a PureLink™ RNA Mini Kit with on-column DNase treatment (Invitrogen™, ThermoFisher Scientific, Sydney, Australia) according to the manufacturer's instructions. cDNA was synthesized from 200 ng RNA using a First Strand cDNA Synthesis Kit (ThermoFisher Scientific, Sydney, Australia). Quantitative PCR (qPCR) reactions were prepared in a total volume of 10 μL with PowerUp SYBR Green Master Mix (ThermoFisher Scientific, Sydney, Australia) and 0.1 mM forward and reverse primers. StepOnePlus PCR equipment (Applied Biosystems) was used to run the samples, with 2 min at 95 °C, then 40 cycles of 3 s at 95 °C and 30 s at 60 °C, followed by a melt curve. Relative mRNA expression was analyzed using the $2^{-\Delta CT}$ method, after being normalized with two housekeeping genes (18s rRNA and GAPDH).

Table 1. Primers used in this study.

	Forward Primer (5′-3′)	Reverse Primer (5′-3′)
ALP	TCAGAAGCTAACACCAACG	TTGTACGTCTTGGAGAGGGC
OPN	TCACCTGTGCCATACCAGTTAA	TGAGATGGGTCAGGGTTTAGC
RUNX2	GGAGTGGACGAGGCAAGAGTTT	AGCTTCTGTCTGTGCCTTCTGG
OSX	GCAAAGCAGGCACAAAGAAG	CAGGTGAAAGGAGCCCATTAG
BSP	AGGCTGAGAATACCACACTTTC	GGATTGCAGCTAACCCTGTAT
18s	TTCGGAACTGAGGCCATGAT	CGAACCTCCGACTTCGTTC
GAPDH	TCAGCAATGCATCCTGCAC	TCTGGGTGGCAGTGATGGC

2.7. Alizarin Red S Staining

For the differentiation assay, cells were seeded at a density of 10,000 cells per 1×1 cm^2 fibrous substrates for three weeks before Alizarin Red S staining for calcium mineral formation from three replicates. Samples were incubated with a 40 mM Alizarin Red S solution (pH 4.1) for 15 min and images were taken using a Nikon microscope. Then, 10% of cetylpyridinium chloride (200 µL) was used to dissolve the staining and the absorbance was measured at 540 nm in an Infinite Pro spectrophotometer.

The DNA content of each sample was measured using a PicoGreen dsDNA quantitation kit (Invitrogen), according to our previous work [28]. Alizarin Red S absorbance was normalized with DNA content from each well to further measure the relative calcium content.

2.8. Statistical Analysis

All data are displayed as the mean $\pm$ standard deviation (SD). The statistical differences between TCP, PCL-R, and PCL-A were determined in GraphPad Prism 9.0 software (GraphPad Software, San Diego, CA, USA) using a one-way analysis of variance (ANOVA) followed by Tukey's multiple comparison tests. A $p < 0.05$ was considered statically significant.

3. Result and Discussion

3.1. Focal Adhesions Are Altered on Aligned Fibers on a PCL Porous Scaffold

The majority of the current literature has investigated the effect of 2D biomaterial cues (such as topography) on understanding cell behaviour via mechanosensing and epigenetics on a given cell type [35–37]. There is limited knowledge of how fiber alignment regulates cellular/nuclear mechanosensing and epigenetics in osteoblasts. The present study successfully fabricated a PCL fibrous scaffold with aligned and random 3D topographical microenvironments for osteoblast cellular and nuclear mechanics by the additive manufacturing, MEW, fabrication technique.

SEM images (Figure 1a) demonstrate the 3D PCL fibrous scaffolds with random (PCL-R) and aligned (PCL-A) fibers fabricated by MEW. As the fabrication parameters were consistent, the fiber diameters of random and aligned fibers did not show any significant changes ($p < 0.05$). The PCL-R diameter size was 19.9 ± 4.1 µm, whereas PCL-A was 18.1 ± 2.5 µm (Figure 1b). The pore size (distance between fibers from center to center) was comparable between both random and aligned fibers (PCL-R: 207 ± 39 µm; PCL-A: 202 ± 15 µm). The obtained fiber diameters results were in good agreement with our previous work [28]; however, the PCL-A group's pore size was smaller than ours [28]. This may be caused by technical variations during different manufacturing batches, while our PCL-R and PCL-A had similar pore sizes (~200 µm) since scaffold pore size is one of the critical factors directing adhesion and differentiation [38,39]. Additionally, mechanical testing demonstrated that the randomly organized membranes were softer than the aligned fibers with a respective tensile modulus of 32.5 ± 0.4 kPa and 56.7 ± 22.5 kPa, respectively. Aside from the osteoblasts investigated in the current study, these 3D platforms can be applied to other cell types for mechanotransduction and epigenetics research.

Given that cell morphology drastically changes in a 3D microenvironment vs. 2D substrate, the question arises as to whether the molecular mechanisms of 3D mechanotransduction also differ. The architecture of adhesion complexes in 3D matrices is remarkably different from those of cells in 2D TCP cultures [40], which is in line with the results in the current study. Although nano-scaled FAs (i.e., a key FA protein and 'molecular clutch'–vinculin [41]) composition is comparable between 2D and 3D microenvironment, actin fibers and FAs in 3D are reduced and smaller compared to the 2D situation [42]. The primary hOBs were cultured on 2D (tissue culture plate-TCP) and 3D geometries (random and aligned PCL porous membranes coated with fetal bovine serum) for 24 h. The data demonstrated a significant cell shape difference between the groups (spread on PCL-R and TCP groups, elongated on PCL-A group). The FAs in hOBs on 3D (both random and aligned) porous scaffolds were mainly detected at the corners or sharp angles formed by adjacent melt electrospun written fibers (yellow arrows), with significantly reduced vinculin (a key focal adhesion protein) numbers (Figure 2b), fewer FAs containing F-actin fibers (Figure 2c), and mature vinculin (>10 μm; Figure 2d) when compared to the 2D TCP substrate. Furthermore, there were more FAs numbers and less mature FAs in hOBs on the random porous membranes than the aligned porous membranes (Figure 2b,d).

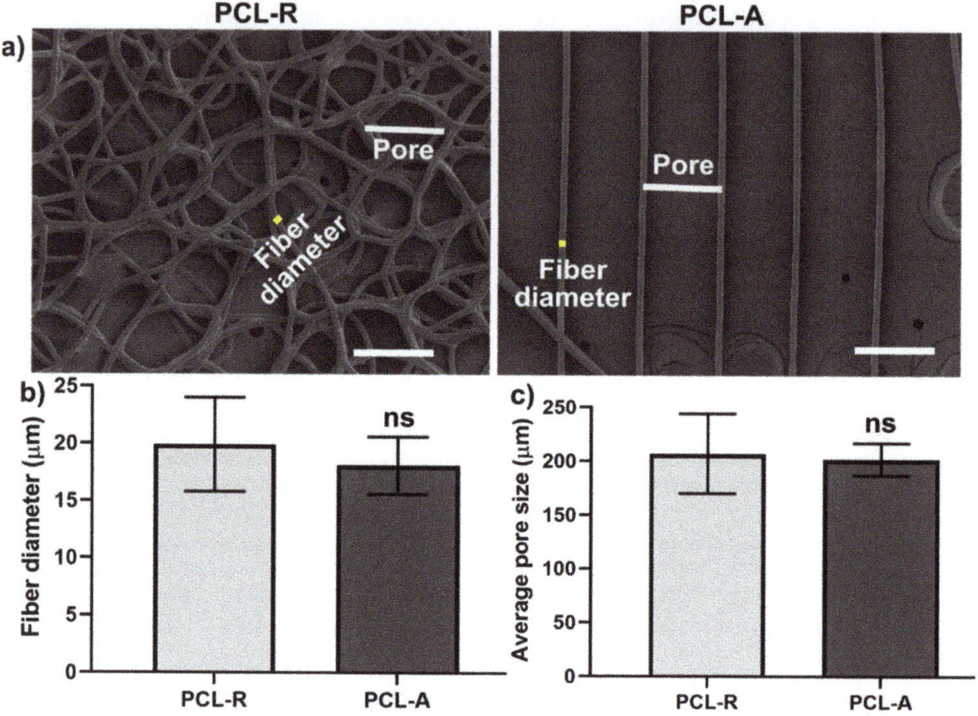

Figure 1. SEM images (**a**), fiber and pore size quantification (**b**,**c**) of electrospun 3D polycaprolactone (PCL) porous scaffold with different geometries using the melt electrospinning direct writing mode. Scale bar in a: 200 μm. ns: no significant difference between PCL-R and PCL-A groups.

Our data demonstrated that the 3D microenvironment (for both random and aligned PCL porous scaffold) led to significantly reduced numbers and smaller vinculin FAs in human osteoblasts compared to 2D TCP (Figure 2), which was in agreement with previous reports although using 3D hydrogels [40,42,43]. Whereas FAs adopted a spindle shape on the 2D TCP, our data showed that 3D FAs were round-shaped (Figure 2a) and were mainly located in the vicinity of optus angles between the PLC fibers, which is consistent with data from 2D micropatterning [44]. Fewer FAs numbers (Figure 2b) and elongated nucleus (Figure 2a) were detected in the aligned PCL porous membranes, given that the cell and nucleus need to accommodate the change in cell shape, which is in accordance with cell shape in aligned nano-PCL fibers [22,44,45]. It is worth noting that cells may interact with multiple fibers at the corner of the PCL-A scaffold, while most of the cells might adhere to a single fiber in the PCL-A group with less surface area than random fibers where cells can interact with more than one fiber.

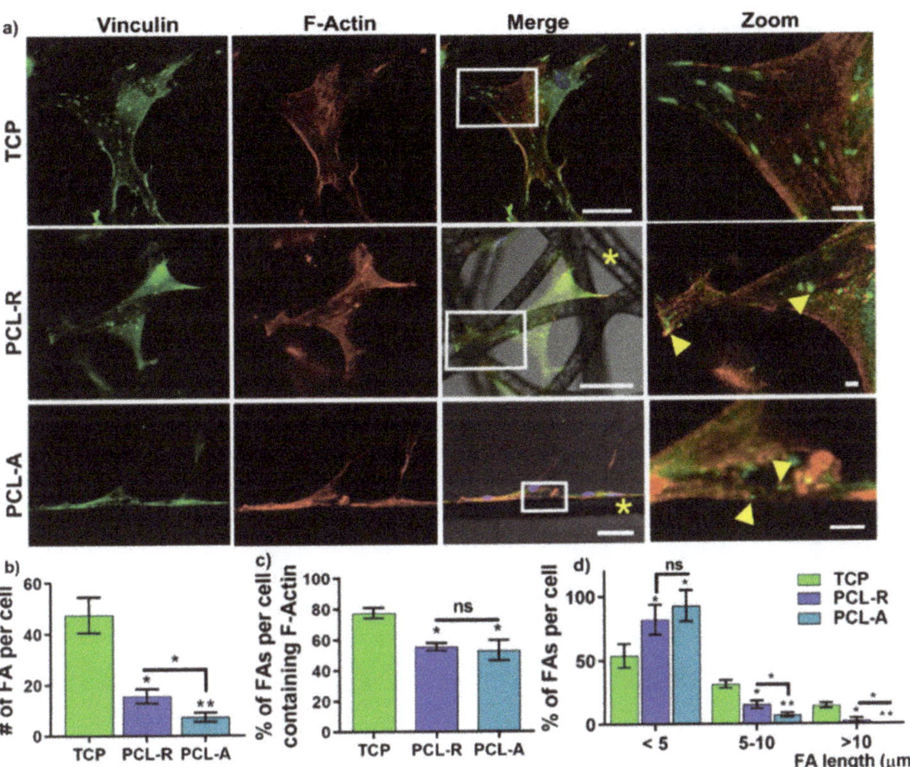

Figure 2. Osteoblast cell shape and focal adhesion on the 3D PCL groups and 2D TCP at 24 h. Representative confocal images (**a**) and quantification (**b–d**) of vinculin in hOBs (>15 cells per group per donor). * PCL fibers; arrows: focal adhesions (FAs); Scale: 50 µm in low-magnification, 10 µm in zoom. * $p < 0.05$, ** $p < 0.02$.

3.2. The Effect of Fiber Alignment on Nucleus Mechanosensing

Besides the cellular mechanosensing, the nucleus is mechanosensitive and linked mechanically to the extracellular matrix via the cytoskeleton that transmits forces to the nuclear envelope (via LAMIN A/C protein) and mediates the cell's genome and epigenome [12,46]. Our data demonstrate that LAMIN A/C was detected in both 2D and 3D, although the aligned fibers led to an elongated nucleus with enhanced LAMIN A/C fluorescence (Figure 3a). YAP is a known key mechanotransductive transcription factor that contributes to the retention storage of the mechanical 'memory' of past cell–matrix interactions [11]; our previous research showed that in 2D substrates, higher vinculin expression led to higher nuclear YAP accumulation and nuclear mechanical tension at an early point (24 h) and was associated with increased osteogenic differentiation in human mesenchymal stem cells [8,45]. LAMIN A/C and YAP were stained for hOBs on the 3D porous scaffolds and TCP (Figure 3). Higher nuclear LAMIN A/C expression with a significantly smaller, elongated nucleus was found on the aligned PCL fibrous substrates, where nuclear shape, area, circularity, and aspect ratio were similar between the PCL-R and TCP groups. It was noted that YAP was mainly localized in the nucleus in 2D culture, while there was a significantly reduced nuclear YAP (%) expression in both 3D groups (Figure 3a,e), suggesting that the nucleus is "softer" in the 3D PCL porous membranes compared to the 2D TCP.

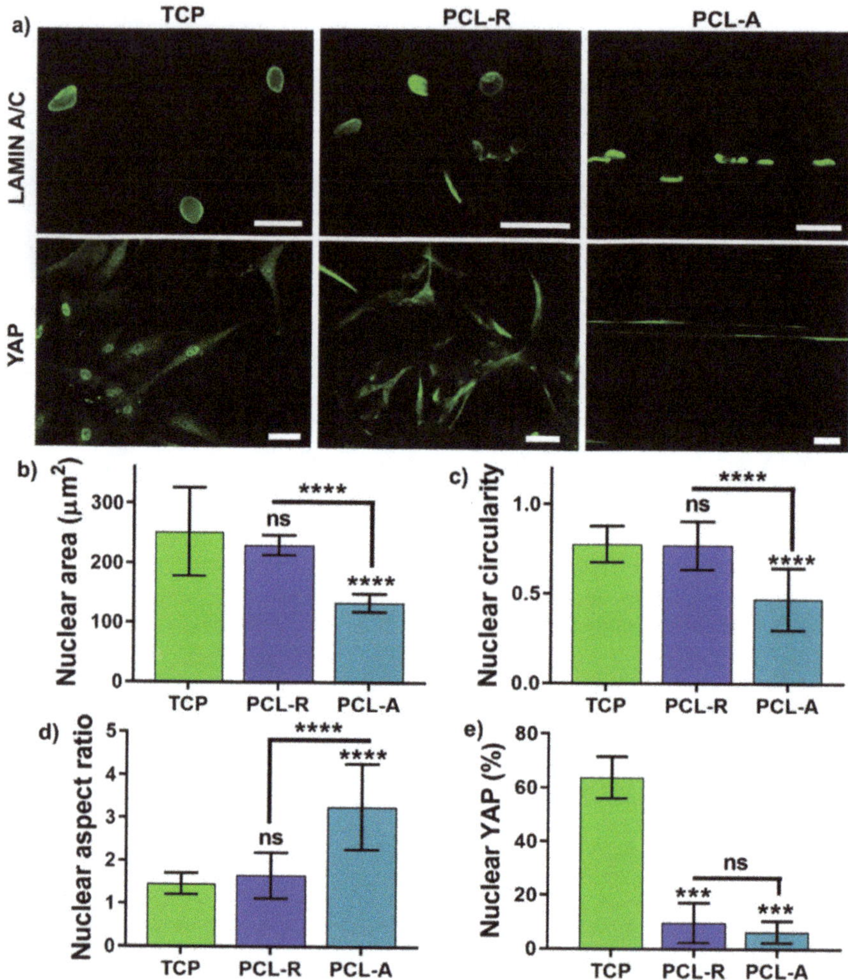

Figure 3. Nuclear LAMIN A/C, yes-associated protein (YAP) expression (**a**) and quantification of nuclear area (**b**), circularity (**c**), aspect ratio (**d**) and nuclear YAP localization (%) (**e**) in hOBs after 24 h. Scale in (**a**): 50 μm. *** $p < 0.001$, ****, $p < 0.0001$.

It has previously been demonstrated that nuclear YAP/TAZ interacts with the TEAD transcription factor and recruits the nucleosome remodelling deacetylase complex on target genes, causing histone deacetylation [47]. However, it is unclear whether YAP localization would alter DNA epigenetics, in terms of global 5mC, 5hmC, and m6dA methylation.

3.3. DNA Epigenetics Is Modulated by Fiber Alignment

To explore the link between nuclear stiffness and global methylation, the effect of different fiber alignment on 5mC, 5hmC, and m6dA expression in hOBs was determined at 24 h. The comparison between random and aligned fibers showed that a mixture of spread and round cell nucleus was observed in the PCL-R group, while the PCL-A group led to a stretched cell with an elongated nucleus (Figure 4a), which is in line with cells cultured on random and aligned chemically cross-linked gelatin fibers in a PCL fibrous substrate [45].

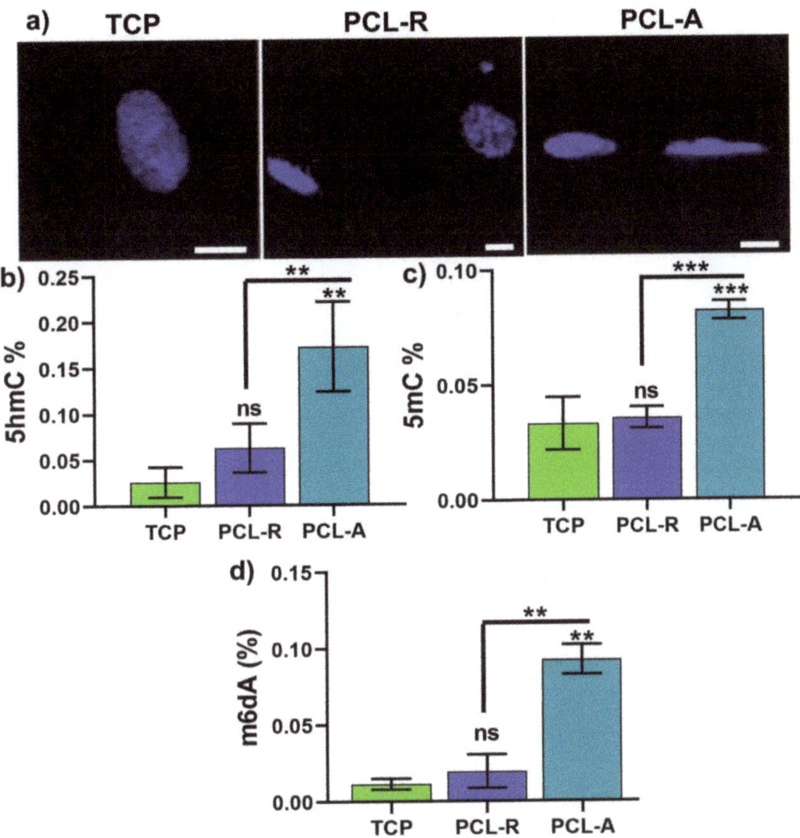

Figure 4. Representative nuclei (**a**), global 5mC (**b**), 5hmC (**c**), and m6dA (**d**) methylation changes in hOBs on different fiber aligned membrane after 24 h. Scale in (**a**): 50 µm. m6dA: N-6 methylated deoxyadenosine; 5mC: 5-Methylcytosine; 5hmC: 5-Hydroxymethylcytosine. Data: mean ± SD. ** $p < 0.02$, *** $p < 0.001$.

Global methylation results indicate no significant difference between TCP and the random group in global 5mC, 5hmC, and m6dA methylation. However, the aligned PCL fibers led to increased global hypermethylation for 5mC (Figure 4b), 5hmC (Figure 4c), and m6dA (Figure 4d) compared to both 2D TCP and PCL-R groups.

3.4. Cell Differentiation on Different Fiber Aligned Porous Membranes

Our previous research showed that in 2D substrates, higher vinculin expression led to higher nuclear YAP accumulation and nuclear mechanical tension at an early point (24 h), with increased long-term (3-week) osteogenic differentiation in human mesenchymal stem cells [8,45]. Our data showed that 3D PCL scaffolds led to significantly decreased FAs (Figure 1a) and nuclear YAP localization (Figure 3a,e), compared to the 2D groups. Long-term osteogenic differentiation was performed on PCL-R, PCL-A, and TCP groups for one and three weeks in osteoinductive media. Osteogenic markers (*ALP, OPN, RUNX2, OSX, BSP*) on different substrates were performed by RT-qPCR after 1-week osteogenic differentiation. The results showed that PCL-A group had significantly increased gene expression of osteogenic markers such as *ALP, OSX*, and *BSP* compared to 2D TCP. Moreover, *ALP* and *RUNX2* were significantly enhanced in the PCL-A group in comparison to the PCL-R group. However, in contrast to the gene expression patterns, Alizarin Red

S staining showed that all three groups enabled the osteogenic differentiation in hOBs (Figure 5b,c) and the quantification of the ARS normalized by DNA content did not result in any significant difference (Figure 5d).

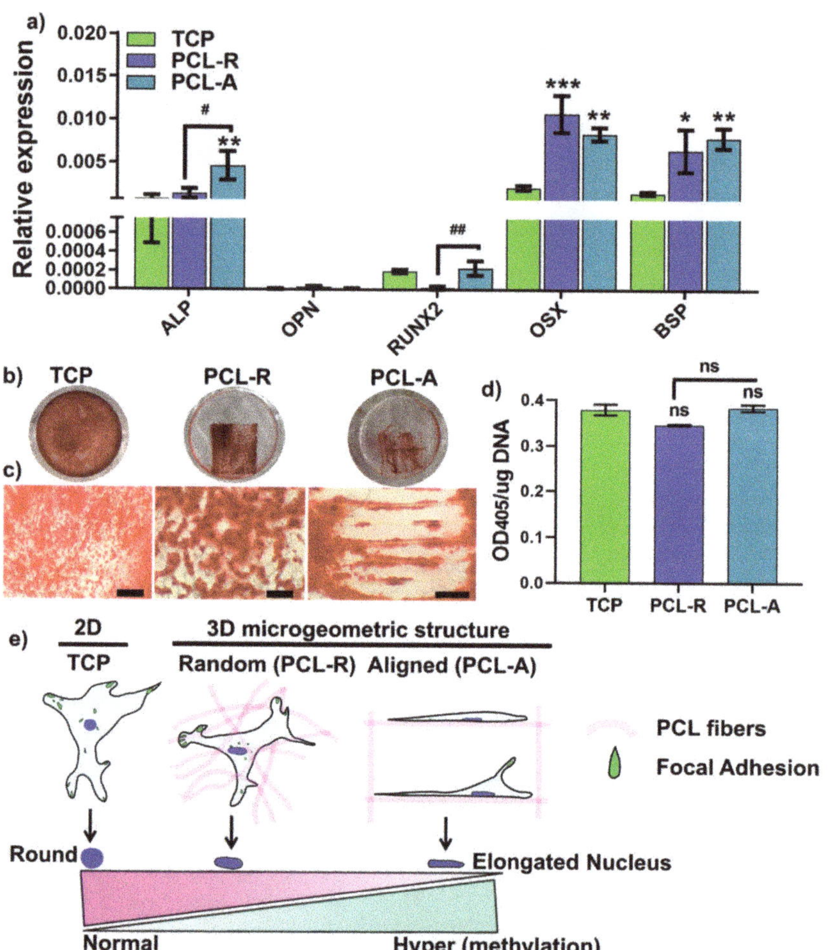

Figure 5. Gene expression of osteogenic markers (**a**) after 1-week osteogenic culture. * $p < 0.05$, ** $p < 0.002$; *** $p < 0.0002$ vs. TCP, # $p < 0.05$, ## $p < 0.002$ between PCL-R and PCL-A. Alizarin Red staining (**b,c**) and quantification (**d**) of hOBs after 3-week osteogenic differentiation in random or aligned PCL porous scaffolds. Scale for image (**b**): 50 μm. ns: no significant difference. (**e**) Schematic illustration of osteoblast response to fiber alignment via altered focal adhesion, nuclear mechanosensing, and global methylation.

In 2D substrates, it is well-known that higher nuclear YAP and FAs are associated with enhanced osteogenic differentiation and mineralization in MSCs [8], but it is unclear whether this applies to 3D microenvironments in osteoblasts. Our results showed a different trend that lower nuclear YAP and fewer FAs (24 h post-seeding) in the PCL-A group led to higher gene expression of early osteogenic markers (7-day post osteogenic differentiation), with no change in calcium mineralization after 3-week osteogenic induction. This contrary trend between mechanotransduction and osteogenic differentiation may explain that osteogenic differentiation and mineralization may be modulated by other complex mechanisms, which requires more studies to explore the underlying mechanisms.

The study has its limitations: (1) different fiber (nano- and micro-scaled) and pore size (20–200 µm) of 3D mPCL scaffolds were not included to compare cell behaviour; and (2) more assays are required to confirm the osteogenic capability of 3D PCL scaffolds. Future studies with various fiber diameters, pore sizes, and more differentiation assays should be considered.

Taken together, the results indicate that aligned fibers within 3D PCL fibrous substrates lead to reduced nano-scaled focal adhesion and nuclear YAP localization compared to 2D TCP. Furthermore, aligned 3D fibers lead to an elongated cell and nuclear shape, resulting in hypermethylation of 5mC, 5hmC, and m6dA compared to random fibers. This altered cellular/nuclear mechanosensing and DNA epigenetics may provide new insights into the endpoint differentiation outcome between 3D random and aligned porous membranes (Figure 5e). However, more studies using cutting-edge techniques such as methylated DNA immunoprecipitation (MeDIP) next-generation sequencing are required for investigating the whole methylome signature. Hence, changing the local microenvironment via different fiber alignment may be an effective strategy for altering nuclear stiffness and epigenetics in hOBs.

4. Conclusions

Although there is still much unknown about how biophysical cues of random and aligned 3D fibrous substrates can impact nuclear signaling, the present study may advance our understanding of random and aligned fibers modulating the epigenetic regulation of osteoblast cells through mechanotransduction. In addition, our findings suggest that fiber alignment acts as an epigenetic factor that affects cell and nuclear mechanosensing, and extended exposure to defined geometric microenvironments may lead to changes in the hOBs phenotype and fate determination of osteoblast cells. Our results highlight the importance of understanding how 3D geometric physical aspects with random and aligned fiber physical cues can alter cellular/nuclear nano-mechanics and DNA epigenetics. Future studies using other cell types are required to determine how 3D mechanotransduction may be harnessed to develop novel biomaterials for tissue engineering and regenerative medicine.

Author Contributions: Conceptualization: P.H. and S.I.; Methodology: P.H., C.V. and A.A.-h.; Writing–original draft: P.H.; Writing–review & editing: P.H., C.V., A.A.-h. and S.I. All authors have read and agreed to the published version of the manuscript.

Funding: This work was supported by the Australian Dental Research Foundation (grant number: 534-2019) and UQ early career research grant (grant number: UQECR1946153).

Institutional Review Board Statement: The study was conducted according to the guidelines of the Declaration of Helsinki, and approved by the Human Ethics Committee of the University of Queensland (Approval number: 2019000134 and date of approval: 01/02/2019).

Informed Consent Statement: Informed consent was obtained from all subjects involved in the study.

Data Availability Statement: The data presented in this study are openly available online.

Conflicts of Interest: The authors declare no conflict of interest.

References

1. Lv, L.W.; Tang, Y.M.; Zhang, P.; Liu, Y.S.; Bai, X.S.; Zhou, Y.S. Biomaterial Cues Regulate Epigenetic State and Cell Functions—A Systematic Review. *Tissue Eng. Part B Rev.* **2018**, *24*, 112–132. [CrossRef] [PubMed]
2. Matsugaki, A.; Aramoto, G.; Ninomiya, T.; Sawada, H.; Hata, S.; Nakano, T. Abnormal arrangement of a collagen/apatite extracellular matrix orthogonal to osteoblast alignment is constructed by a nanoscale periodic surface structure. *Biomaterials* **2015**, *37*, 134–143. [CrossRef]
3. Nakanishi, Y.; Matsugaki, A.; Kawahara, K.; Ninomiya, T.; Sawada, H.; Nakano, T. Unique arrangement of bone matrix orthogonal to osteoblast alignment controlled by Tspan11-mediated focal adhesion assembly. *Biomaterials* **2019**, *209*, 103–110. [CrossRef]
4. Kubow, K.E.; Conrad, S.K.; Horwitz, A.R. Matrix microarchitecture and myosin II determine adhesion in 3D matrices. *Curr. Biol.* **2013**, *23*, 1607–1619. [CrossRef]
5. Ma, C.; Kuzma, M.L.; Bai, X.; Yang, J. Biomaterial-Based Metabolic Regulation in Regenerative Engineering. *Adv. Sci.* **2019**, *6*, 1900819. [CrossRef] [PubMed]

6. Amani, H.; Arzaghi, H.; Bayandori, M.; Dezfuli, A.S.; Pazoki-Toroudi, H.; Shafiee, A.; Moradi, L. Controlling Cell Behavior through the Design of Biomaterial Surfaces: A Focus on Surface Modification Techniques. *Adv. Mater. Interfaces* **2019**, *6*, 1900572. [CrossRef]

7. Hynes, R.O. The extracellular matrix: Not just pretty fibrils. *Science* **2009**, *326*, 1216–1219. [CrossRef]

8. Han, P.; Frith, J.E.; Gomez, G.A.; Yap, A.S.; O'Neill, G.M.; Cooper-White, J.J. Five Piconewtons: The Difference between Osteogenic and Adipogenic Fate Choice in Human Mesenchymal Stem Cells. *ACS Nano* **2019**, *13*, 11129–11143. [CrossRef] [PubMed]

9. Haupt, A.; Minc, N. How cells sense their own shape—Mechanisms to probe cell geometry and their implications in cellular organization and function. *J. Cell Sci.* **2018**, *131*, jcs214015. [CrossRef]

10. Swaminathan, V.; Waterman, C.M. The molecular clutch model for mechanotransduction evolves. *Nat. Cell Biol.* **2016**, *18*, 459–461. [CrossRef] [PubMed]

11. Dupont, S.; Morsut, L.; Aragona, M.; Enzo, E.; Giulitti, S.; Cordenonsi, M.; Zanconato, F.; Le Digabel, J.; Forcato, M.; Bicciato, S.; et al. Role of YAP/TAZ in mechanotransduction. *Nature* **2011**, *474*, 179–183. [CrossRef] [PubMed]

12. Cho, S.; Irianto, J.; Discher, D.E. Mechanosensing by the nucleus: From pathways to scaling relationships. *J. Cell Biol.* **2017**, *216*, 305–315. [CrossRef] [PubMed]

13. Crowder, S.W.; Leonardo, V.; Whittaker, T.; Papathanasiou, P.; Stevens, M.M. Material Cues as Potent Regulators of Epigenetics and Stem Cell Function. *Cell Stem Cell* **2016**, *18*, 39–52. [CrossRef] [PubMed]

14. Worley, K.; Certo, A.; Wan, L.Q. Geometry-Force Control of Stem Cell Fate. *Bionanoscience* **2013**, *3*, 43–51. [CrossRef]

15. Larsson, L. Current Concepts of Epigenetics and Its Role in Periodontitis. *Curr. Oral Health Rep.* **2017**, *4*, 286–293. [CrossRef] [PubMed]

16. Bird, A. DNA methylation patterns and epigenetic memory. *Genes Dev.* **2002**, *16*, 6–21. [CrossRef]

17. Robertson, K.D.; Wolffe, A.P. DNA methylation in health and disease. *Nat. Rev. Genet.* **2000**, *1*, 11–19. [CrossRef] [PubMed]

18. Ai, T.; Zhang, J.; Wang, X.; Zheng, X.; Qin, X.; Zhang, Q.; Li, W.; Hu, W.; Lin, J.; Chen, F. DNA methylation profile is associated with the osteogenic potential of three distinct human odontogenic stem cells. *Signal Transduct. Target. Ther.* **2018**, *3*, 1. [CrossRef]

19. Xin, T.-Y.; Yu, T.-T.; Yang, R.-L. DNA methylation and demethylation link the properties of mesenchymal stem cells: Regeneration and immunomodulation. *World J. Stem Cells* **2020**, *12*, 351–358. [CrossRef] [PubMed]

20. Han, P.; Bartold, P.M.; Salomon, C.; Ivanovski, S. Salivary Outer Membrane Vesicles and DNA Methylation of Small Extracellular Vesicles as Biomarkers for Periodontal Status: A Pilot Study. *Int. J. Mol. Sci.* **2021**, *22*, 2423. [CrossRef] [PubMed]

21. Larsson, L.; Pilipchuk, S.P.; Giannobile, W.V.; Castilho, R.M. When epigenetics meets bioengineering-A material characteristics and surface topography perspective. *J. Biomed. Mater. Res. B Appl. Biomater.* **2018**, *106*, 2065–2071. [CrossRef] [PubMed]

22. Downing, T.L.; Soto, J.; Morez, C.; Houssin, T.; Fritz, A.; Yuan, F.; Chu, J.; Patel, S.; Schaffer, D.V.; Li, S. Biophysical regulation of epigenetic state and cell reprogramming. *Nat. Mater.* **2013**, *12*, 1154–1162. [CrossRef] [PubMed]

23. Abbasi, N.; Abdal-hay, A.; Hamlet, S.; Graham, E.; Ivanovski, S. Effects of Gradient and Offset Architectures on the Mechanical and Biological Properties of 3-D Melt Electrowritten (MEW) Scaffolds. *ACS Biomater. Sci. Eng.* **2019**, *5*, 3448–3461. [CrossRef] [PubMed]

24. Abdal-hay, A.; Abbasi, N.; Gwiazda, M.; Hamlet, S.; Ivanovski, S. Novel polycaprolactone/hydroxyapatite nanocomposite fibrous scaffolds by direct melt-electrospinning writing. *Eur. Polym. J.* **2018**, *105*, 257–264. [CrossRef]

25. Barrows, T. Degradable implant materials: A review of synthetic absorbable polymers and their applications. *Clin. Mater.* **1986**, *1*, 233–257. [CrossRef]

26. Abdal-hay, A.; Raveendran, N.T.; Fournier, B.; Ivanovski, S. Fabrication of biocompatible and bioabsorbable polycaprolactone/magnesium hydroxide 3D printed scaffolds: Degradation and in vitro osteoblasts interactions. *Compos. Part B Eng.* **2020**, *197*, 108158. [CrossRef]

27. Abdal-hay, A.; Amna, T.; Lim, J.K. Biocorrosion and Osteoconductivity of PCL/nHAp Composite Porous Film-Based Coating of Magnesium Alloy. *Solid State Sci.* **2012**, *18*, 131–140. [CrossRef]

28. Gwiazda, M.; Kumar, S.; Swieszkowski, W.; Ivanovski, S.; Vaquette, C. The effect of melt electrospun writing fiber orientation onto cellular organization and mechanical properties for application in Anterior Cruciate Ligament tissue engineering. *J. Mech. Behav. Biomed. Mater.* **2020**, *104*, 103631. [CrossRef] [PubMed]

29. Dan, H.; Vaquette, C.; Fisher, A.G.; Hamlet, S.M.; Xiao, Y.; Hutmacher, D.W.; Ivanovski, S. The influence of cellular source on periodontal regeneration using calcium phosphate coated polycaprolactone scaffold supported cell sheets. *Biomaterials* **2014**, *35*, 113–122. [CrossRef]

30. Haase, H.R.; Ivanovski, S.; Waters, M.J.; Bartold, P.M. Growth hormone regulates osteogenic marker mRNA expression in human periodontal fibroblasts and alveolar bone-derived cells. *J. Periodontal Res.* **2003**, *38*, 366–374. [CrossRef] [PubMed]

31. Han, P.; Ivanovski, S.; Crawford, R.; Xiao, Y. Activation of the Canonical Wnt Signaling Pathway Induces Cementum Regeneration. *J. Bone Miner. Res.* **2015**, *30*, 1160–1174. [CrossRef] [PubMed]

32. Blaudez, F.; Ivanovski, S.; Ipe, D.; Vaquette, C. A comprehensive comparison of cell seeding methods using highly porous melt electrowriting scaffolds. *Mater. Sci. Eng. C Mater.* **2020**, *117*, 111282. [CrossRef] [PubMed]

33. Zhou, Y.; Han, S.; Xiao, L.; Han, P.; Wang, S.; He, J.; Chang, J.; Wu, C.; Xiao, Y. Accelerated host angiogenesis and immune responses by ion release from mesoporous bioactive glass. *J. Mater. Chem. B* **2018**, *6*, 3274–3284. [CrossRef] [PubMed]

34. Han, P.; Lloyd, T.; Chen, Z.; Xiao, Y. Proinflammatory Cytokines Regulate Cementogenic Differentiation of Periodontal Ligament Cells by Wnt/Ca(2+) Signaling Pathway. *J. Interferon Cytokine Res.* **2016**, *36*, 328–337. [CrossRef]

35. Cui, Y.; Xiao, Z.; Chen, T.; Wei, J.; Chen, L.; Liu, L.; Chen, B.; Wang, X.; Li, X.; Dai, J. The miR-7 identified from collagen biomaterial-based three-dimensional cultured cells regulates neural stem cell differentiation. *Stem Cells Dev.* **2014**, *23*, 393–405. [CrossRef] [PubMed]

36. Le Beyec, J.; Xu, R.; Lee, S.Y.; Nelson, C.M.; Rizki, A.; Alcaraz, J.; Bissell, M.J. Cell shape regulates global histone acetylation in human mammary epithelial cells. *Exp. Cell Res.* **2007**, *313*, 3066–3075. [CrossRef] [PubMed]

37. Doyle, A.D.; Carvajal, N.; Jin, A.; Matsumoto, K.; Yamada, K.M. Local 3D matrix microenvironment regulates cell migration through spatiotemporal dynamics of contractility-dependent adhesions. *Nat. Commun.* **2015**, *6*, 8720. [CrossRef] [PubMed]

38. Han, Y.; Lian, M.; Wu, Q.; Qiao, Z.; Sun, B.; Dai, K. Effect of Pore Size on Cell Behavior Using Melt Electrowritten Scaffolds. *Front. Bioeng. Biotechnol.* **2021**, *9*, 495. [CrossRef] [PubMed]

39. Lanaro, M.; McLaughlin, M.P.; Simpson, M.J.; Buenzli, P.R.; Wong, C.S.; Allenby, M.C.; Woodruff, M.A. A quantitative analysis of cell bridging kinetics on a scaffold using computer vision algorithms. *Acta Biomater.* **2021**. [CrossRef] [PubMed]

40. Cukierman, E.; Pankov, R.; Stevens, D.R.; Yamada, K.M. Taking cell-matrix adhesions to the third dimension. *Science* **2001**, *294*, 1708–1712. [CrossRef] [PubMed]

41. Grashoff, C.; Hoffman, B.D.; Brenner, M.D.; Zhou, R.; Parsons, M.; Yang, M.T.; McLean, M.A.; Sligar, S.G.; Chen, C.S.; Ha, T.; et al. Measuring mechanical tension across vinculin reveals regulation of focal adhesion dynamics. *Nature* **2010**, *466*, 263–266. [CrossRef] [PubMed]

42. Hakkinen, K.M.; Harunaga, J.S.; Doyle, A.D.; Yamada, K.M. Direct comparisons of the morphology, migration, cell adhesions, and actin cytoskeleton of fibroblasts in four different three-dimensional extracellular matrices. *Tissue Eng. Part A* **2011**, *17*, 713–724. [CrossRef] [PubMed]

43. Fraley, S.I.; Feng, Y.; Krishnamurthy, R.; Kim, D.H.; Celedon, A.; Longmore, G.D.; Wirtz, D. A distinctive role for focal adhesion proteins in three-dimensional cell motility. *Nat. Cell Biol.* **2010**, *12*, 598–604. [CrossRef] [PubMed]

44. Thery, M.; Racine, V.; Piel, M.; Pepin, A.; Dimitrov, A.; Chen, Y.; Sibarita, J.B.; Bornens, M. Anisotropy of cell adhesive microenvironment governs cell internal organization and orientation of polarity. *Proc. Natl. Acad. Sci. USA* **2006**, *103*, 19771–19776. [CrossRef] [PubMed]

45. Wang, M.; Cui, C.; Ibrahim, M.M.; Han, B.; Li, Q.; Pacifici, M.; Lawrence, J.T.R.; Han, L.; Han, L.-H. Regulating Mechanotransduction in Three Dimensions using Sub-Cellular Scale, Crosslinkable Fibers of Controlled Diameter, Stiffness, and Alignment. *Adv. Funct. Mater.* **2019**, *29*, 1808967. [CrossRef]

46. Isermann, P.; Lammerding, J. Nuclear mechanics and mechanotransduction in health and disease. *Curr. Biol.* **2013**, *23*, R1113–R1121. [CrossRef] [PubMed]

47. Kim, M.; Kim, T.; Johnson, R.L.; Lim, D.S. Transcriptional co-repressor function of the hippo pathway transducers YAP and TAZ. *Cell Rep.* **2015**, *11*, 270–282. [CrossRef] [PubMed]

 *nanomaterials*

Article

Alkali-Treated Titanium Coated with a Polyurethane, Magnesium and Hydroxyapatite Composite for Bone Tissue Engineering

Mahmoud Agour [1,†], Abdalla Abdal-hay [2,3,*,†], Mohamed K. Hassan [1,4], Michal Bartnikowski [2] and Sašo Ivanovski [2,*]

1 Department of Production Engineering and Design, Faculty of Engineering, Minia University, Minia 61112, Egypt; agour.mahmoud@gmail.com (M.A.); mkibrahiem@uqu.edu.sa (M.K.H.)
2 Centre for Orofacial Regeneration, Reconstruction and Rehabilitation (COR3), School of Dentistry, Herston Campus, The University of Queensland, 288 Herston Road, Herston, QLD 4006, Australia; michal.bartnikowski@uq.edu.au
3 Department of Engineering Materials and Mechanical Design, Faculty of Engineering, South Valley University, Qena 83523, Egypt
4 Department of Mechanical Engineering, College of Engineering, Umm Al-Qura University (UQU), Mecca 24381, Saudi Arabia
* Correspondence: abdalla.ali@uq.edu.au (A.A.-h.); s.ivanovski@uq.edu.au (S.I.)
† Those authors contributed equally to this work.

Abstract: The aim of this study was to form a functional layer on the surface of titanium (Ti) implants to enhance their bioactivity. Layers of polyurethane (PU), containing hydroxyapatite (HAp) nanoparticles (NPs) and magnesium (Mg) particles, were deposited on alkali-treated Ti surfaces using a cost-effective dip-coating approach. The coatings were assessed in terms of morphology, chemical composition, adhesion strength, interfacial bonding, and thermal properties. Additionally, cell response to the variably coated Ti substrates was investigated using MC3T3-E1 osteoblast-like cells, including assessment of cell adhesion, cell proliferation, and osteogenic activity through an alkaline phosphatase (ALP) assay. The results showed that the incorporation of HAp NPs enhanced the interfacial bonding between the coating and the alkali-treated Ti surface. Furthermore, the presence of Mg and HAp particles enhanced the surface charge properties as well as cell attachment, proliferation, and differentiation. Our results suggest that the deposition of a bioactive composite layer containing Mg and HAp particles on Ti implants may have the potential to induce bone formation.

Keywords: magnesium and its alloys; hydroxyapatite; surface modifications; titanium implants; corrosion analysis; biocompatibility; bioactivity

Citation: Agour, M.; Abdal-hay, A.; Hassan, M.K.; Bartnikowski, M.; Ivanovski, S. Alkali-Treated Titanium Coated with a Polyurethane, Magnesium and Hydroxyapatite Composite for Bone Tissue Engineering. *Nanomaterials* **2021**, *11*, 1129. https://doi.org/10.3390/nano11051129

Academic Editor: Ilaria Armentano

Received: 25 March 2021
Accepted: 20 April 2021
Published: 27 April 2021

1. Introduction

Titanium (Ti) and its alloys are extensively used in orthopedic implants due to their favorable corrosion resistance, biocompatibility, machinability, and load bearing capability [1–3]. However, because of the bioinert nature of titanium, the osseointegration of Ti implants is relatively slow and can be compromised in sites of limited bone quality and quantity. Various methods of surface modification of Ti implants using physical, chemical, and biochemical treatment methods have been previously established [1,2,4,5], with the aim to enhance osseointegration of the Ti implants and promote their local bioactivity. Previous studies have shown that good surface wettability enhances the attachment, proliferation, and differentiation of osteoblasts and their precursors, leading to enhanced bone healing [4,6–11]. Additionally, surface chemistry plays an important role in the early stages of bone formation, with the greatest benefits derived from osteoconductive materials, such as hydroxyapatite HAp ($Ca_{10}(PO_4)_6(OH)_2$). HAp has been widely used for bone tissue engineering due to its excellent biocompatibility, bioactivity (chemical bonding ability with natural bone), slow in situ biodegradability, high osteoconductivity, osteoinductive potential, non-toxicity, and immunomodulatory properties [12,13]. However, HAp coatings are often problematic due to the low bonding strength of pure HAp coatings and Ti substrates [1,10,14–21]. The difference in the thermal expansion coefficients of Ti and HAp can cause delamination of the coating, leading to the failure of implants clinically.

Recently, magnesium (Mg) alloys or Mg coatings on metallic substrates were introduced in medical applications [1,22,23]. Our recent publication [4] showed that the incorporation of Mg particles into a biodegradable polymer not only enhanced the surface wettability, but also improved cell attachment and proliferation. We thus speculated that a composite polymer film composed of biodegradable metallic Mg particles and HAp NPs, successfully deposited on alkaline-treated Ti surfaces using a simple and cost-effective dip-coating method, is an efficient method to enhance the surface biocompatibility of Ti implants and subsequently improve osseointegration. The incorporation of HAp NPs within the polymer matrix may be capable of enhancing the interfacial strength between the alkali-treated Ti substrate and the polymer matrix by promoting increased hydrogen bonding and inducing superior adhesion. Previous studies have verified the strong interfacial strength between sodium titanate hydrogel layers formed on alkali-treated Ti and HAp particles [24–26]. In addition, HAp NPs present within the polymer matrix can become entrapped in the microporosity of the alkali-treated Ti surface, thus establishing mechanical interlocking with the Ti substrate. Furthermore, the incorporation of Mg particles in a biodegradable polymer matrix is advantageous as the release of acidic byproducts of polymer degradation can be compensated through the increase in pH facilitated by the dissolution of Mg [27–30]. Barrère et al. [31] demonstrated that the existence of Mg in an apatite (a phase of HAp) [13] coating of a titanium surface strengthened the adherence of the coating through the Mg ions, promoting further apatite precipitation.

Therefore, the aim of the present study is to incorporate both Mg metal and HAp NPs into a biodegradable polymer matrix and subsequently dip-coat a thin layer of this composite onto Ti substrates to enhance their bioactivity. Herein, polyurethane polymer (Pu) was utilized as a polymer matrix because it has a moderate degradation rate, good biocompatibility and exhibits a natural self-healing property [32]. In the present study, 5 wt.% synthesized HAp NPs [33] and 10 wt.% Mg microparticles [4,12] were incorporated into a PU polymer matrix before deposition through dip-coating onto alkali-treated Ti substrates. The morphological properties, chemical composition, surface roughness, wettability, adhesion strength, interfacial bonding, and thermal properties of the composite coating layer were investigated. Furthermore, MC3T3-E1 osteoblast-like cell adhesion, proliferation, and alkaline phosphatase (ALP) activity was studied in response to coated and uncoated Ti substrates.

2. Materials and Methods

The present study used a commercially pure titanium (c.p. Ti) sheet (William Gregor Ltd., London, UK), which was cut in a square shape with dimensions of $12 \times 12 \times 2 \ mm^3$. The sample preparation included grinding, polishing, and cleaning, according to previous publications [10,34–38]. Untreated samples were then etched in H_2SO_4:HCl:H_2O = 1:1:1 volume ratio at 60 °C for 1 h to increase the surface roughness [39]. The hydrophilicity was enhanced by subjecting the etched samples to alkaline treatment using the same procedures as described elsewhere [35,36,39]. Briefly, the samples were immersed in 100 mL of an aqueous solution of 5 M NaOH at 60 °C for one day. Before further analysis, the samples were rained three times in distilled water and dried at 50 °C for 24 h at atmospheric conditions [10]. A HAp nanopowder was prepared using a wet chemical procedure, as described previously [4,12,33,40]. Further details on the preparation process of HAp is shown in the Supporting Information. Scanning electron microscope (SEM) image and X-ray Diffraction (XRD) profiles of the synthesized HAp powder are shown in Figure 1a,b. The average diameter of the HAp particles was 120 ± 30 nm.

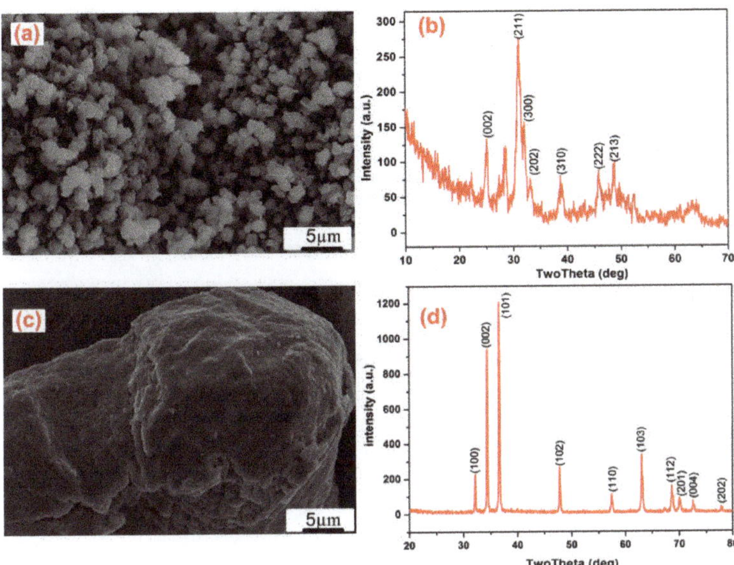

Figure 1. Scanning electron microscopy (SEM) images and X-ray diffraction (XRD) patterns of: (**a**,**b**) HAp, (**c**,**d**) Mg powders.

Coating process: Polymer solution containing 15 g of PU pellets (54,600 Da molecular weight, Estane skythane X595A-11, Lubrizol Advanced Material, Inc., Co., Ltd., Seoul, Korea) was dissolved in 100 mL dimethylformamide (DMF, Loba Chemie PVT. LTD., Mumbai, India). Mg powder with a purity of 99.9% (Kt, Sigma-Aldrich, St. Louis, MO, USA) and particle size range from 40 to 140 µm (SEM morphology and the XRD profile of the as-received Mg particles are shown in Figure 1c,d) was added into the PU solution at 10 wt.% (based on the PU weight contribution). Furthermore, 5 wt.% of HAp NPs, again based on the PU amount, was added into the Mg-particles/PU suspension solution.

Prior to the dip-coating process, all treated and untreated samples were pre-heated over a heating plate at 200 °C for 10 min to remove entrapped air and moisture from the surface. After this preheating process, the samples were immersed in the prepared solutions for about 30 s to facilitate wettability on the surface. Both dipping and withdrawing were performed at the same speed (2.0 mm·s^{-1}) by a dip-coating machine (EF-5100, E-Flex, Bucheon, Korea), as displayed in Figure 2. The coated sheet was then hung in a vacuum oven (10 mbar) at 40 °C for 12 h. The drying temperature was selected based on the glass transition temperature of PU [41,42]. Due to the vertical orientation of the hung sample, there was an outflow of solution from the substrate that could affect the thickness of the coating film. However, previous studies have shown that the coating uniformity is improved when the sample is dried in a vertical rather than horizontal orientation [43–45]. Four groups of samples (S0–S3) were prepared and investigated. Table 1 describes the characteristics of the four groups. Characterization and cell culture experiment are shown in the Supporting Information for further details.

Table 1. Designation of sample groups used in this study.

Sample Designation	Description
S0	Alkaline-treated Ti
S1	Plain PU film coated on alkaline-treated Ti
S2	Mg-particle/PU composite film coated on alkaline-treated Ti
S3	Mg- HAp/PU composite film coated on alkaline-treated Ti

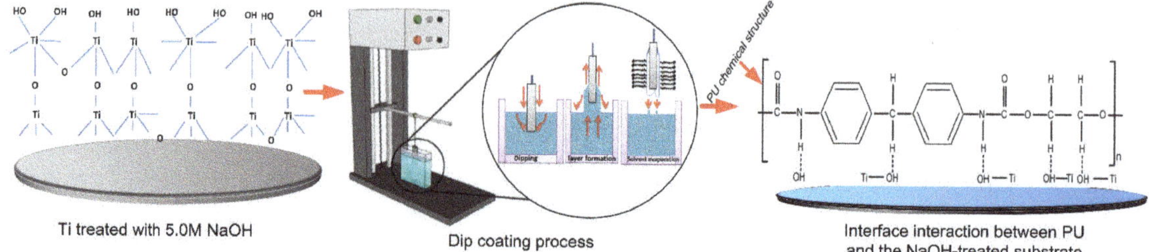

Figure 2. The dip-coating process and the considerable electrostatic intermolecular interaction between the polymer coating and the Ti substrate at the PU polymer side chain. Dotted line indicates electrostatic interaction.

3. Results and Discussion

The XRD curves of the surface of Ti treated with 5 M NaOH solution at 60 °C for 24 h are displayed in Figure 3. A small peak with a broad signal on the 2-theta axis at around 24° was observed. For the S0 group, outstanding peaks due to the α-phase of Ti always existed in the diffraction patterns [46]. Broad, small, and low intensity peaks were detected in the XRD profiles around the two-θ angle of 58°, which was due to the presence of an amorphous sodium titanate hydrogel layer with an atomic ratio similar to the Na_2TiO_3 phase formed by NaOH treatment, as illustrated in previous studies, including ours [33,46,47]. This is likely due to the significant chemical reaction that occurred after the NaOH treatment of the commercially pure Ti surface. However, a previous study [46] did not show the formation of a Na_2TiO_3 gel layer on the surface of Ti, which is in contrast with the present study. This further verifies the well-optimized parameters for NaOH treatment of Ti samples in the present work. The morphological properties and high surface area of the Na_2TiO_3 formed layer (as shown in Figure 4a) are the reason for the obtaining improvement in adhesion of the subsequent PU polymer coating, which will be discussed later. SEM (Figure 4b) and XRD (Figure 3) observations suggested the presence of a new layer of sodium titanate on the surface of S0, which is likely associated with enhanced adhesion performance of the coated films [10]. Figure 5a,b shows the Mg microparticle and HAp nanoparticle distribution within the PU matrix.

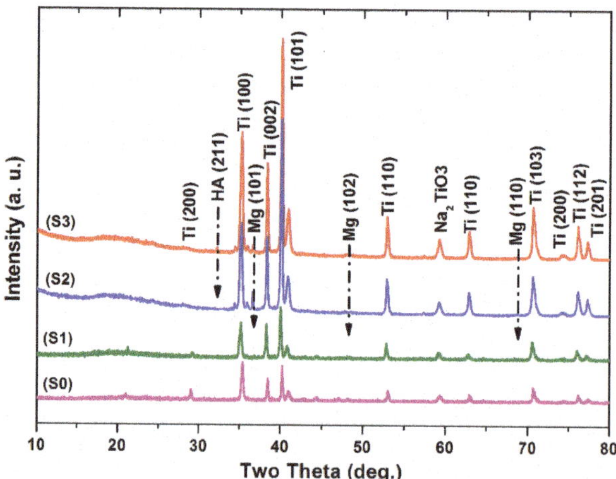

Figure 3. X-ray diffraction profiles of the coated samples. The formation of Na_2TiO_3 gel layer at two-theta with an angle of 58°, incorporation of Mg microparticles and HAp NPs within PU film at 2-theta, 36°, 48°, 57° for Mg particles and 32° for HAp NPs are illustrated by dashed vertical Dotted marks over the XRD profiles.

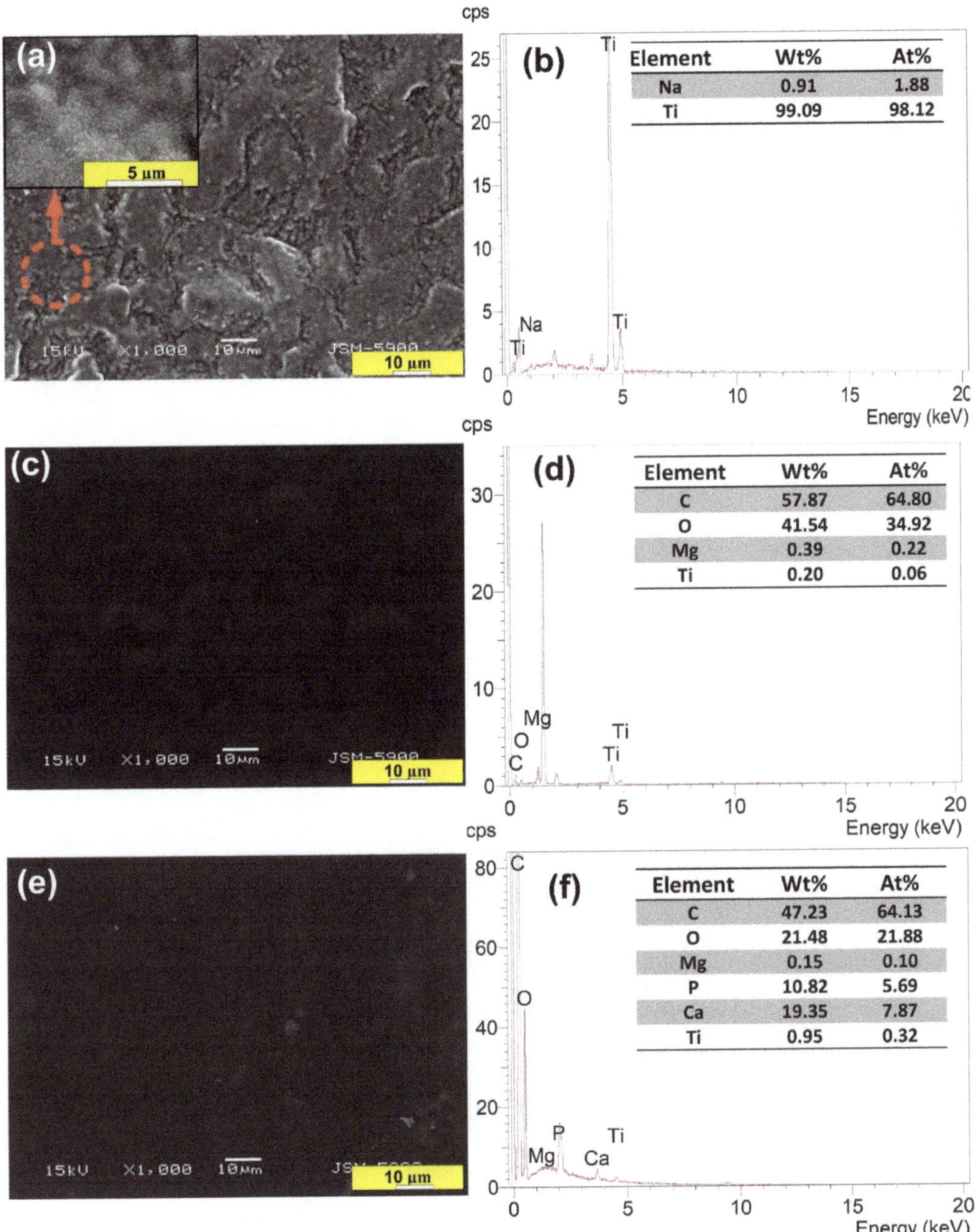

Element	Wt%	At%
Na	0.91	1.88
Ti	99.09	98.12

Element	Wt%	At%
C	57.87	64.80
O	41.54	34.92
Mg	0.39	0.22
Ti	0.20	0.06

Element	Wt%	At%
C	47.23	64.13
O	21.48	21.88
Mg	0.15	0.10
P	10.82	5.69
Ca	19.35	7.87
Ti	0.95	0.32

Figure 4. SEM images and their corresponding EDS profiles of S0 (**a**,**b**), S2 (**c**,**d**), and S3 (**e**,**f**) samples. The analysis of the corresponding SEM image are presented as insets of corresponding panel.

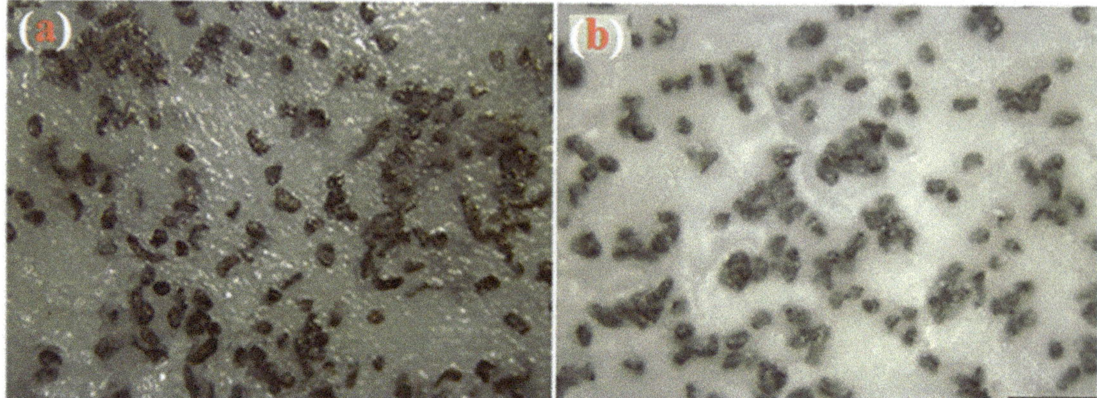

Figure 5. Optical microscopy images showing the distribution of Mg particles (**a**) and Mg-particles/HAp (**b**) throughout the PU layer.

The phase composition of the PU film coated on the untreated and treated Ti surfaces was also investigated using the XRD technique. From the obtained results, it was found that the formation of weak, broad peaks refer to the polymeric phase profile on the Ti surface (see the left-hand side of XRD curves in Figure 3). In addition, the XRD of the composite coated films (S2, S3) showed weak peaks at approximately 34°, 36°, 48°, 57°, indicating the deposition of magnesium particles, and at about 32°, indicating the deposition of HAp particles [30]. SEM images (Figure 4) showed that the S2 group had a smooth and dense morphology, while the S3 group had well-developed upright (columnar) morphological structures, created on the alkaline Ti surface. The surface composition of the coated samples was analyzed by EDS, as depicted in Figure 4d, f. From the EDS analysis, it was confirmed that the Mg (Figure 4d) and HAp particles (Figure 4f) were successfully doped into the PU polymer coating layers. The thickness of the measured coating layer ranged from 8.14 to 9.74 μm, as shown in Figure 6.

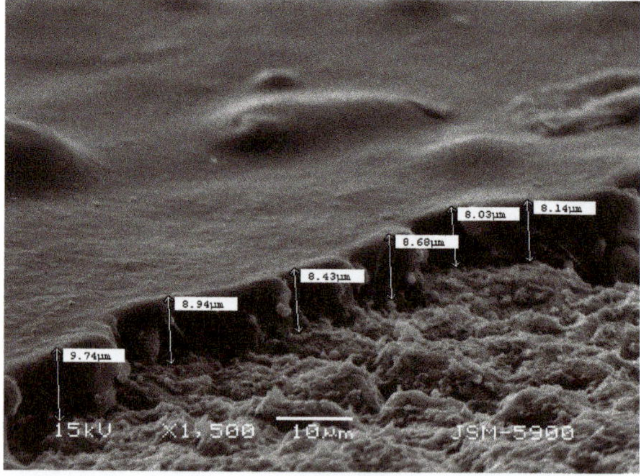

Figure 6. SEM of the cross-section of the composite layer coating on the Ti surfaces, indicating the coating thickness.

The thermal stability of the plain PU coating and the composite coating was investigated using thermogravimetric analysis (TGA), as shown in Figure 6. It appeared that the addition of the Mg particles (Figure 7) improved the thermal stability of the PU as compared to the plain PU (Figure 6, S1). Additionally, the TGA thermogram indicated a similar single step degradation pattern for both S2 and S3. The char yield was higher for S3, which was due to the presence of the inorganic component of HAp. The glass transition temperature for the S3 group, at 50 °C, was marginally higher than that of the S2 group, potentially due to greater secondary interactions and the presence of an additional metal salt. This increase can reasonably be attributed to the coordinated interaction between oxygen and metals/metal ions. The melting point, around 167 °C, appeared to be influenced by the diol component or due to an ordered urea hydrogen bond formation. This may occur due to formation of amine in the presence of residual water/moisture. Finally, the peak around 320 °C indicated degradation. The improvement in degradation temperature in the composite groups is due to the excellent thermal stability of the Mg microparticles capsulated within the PU polymer matrix. In addition, the differential scanning calorimetry (DSC) scans (Figure 7) illustrate a slight shifting of the endothermic peaks and heating enthalpies of the coating groups, which is correlated with the melting of PU hard segments [48]. This might suggest typical interactions between the crystalline regions of PU and the Mg particles that enhance the stabilization of the dynamic thermal properties of the composite films.

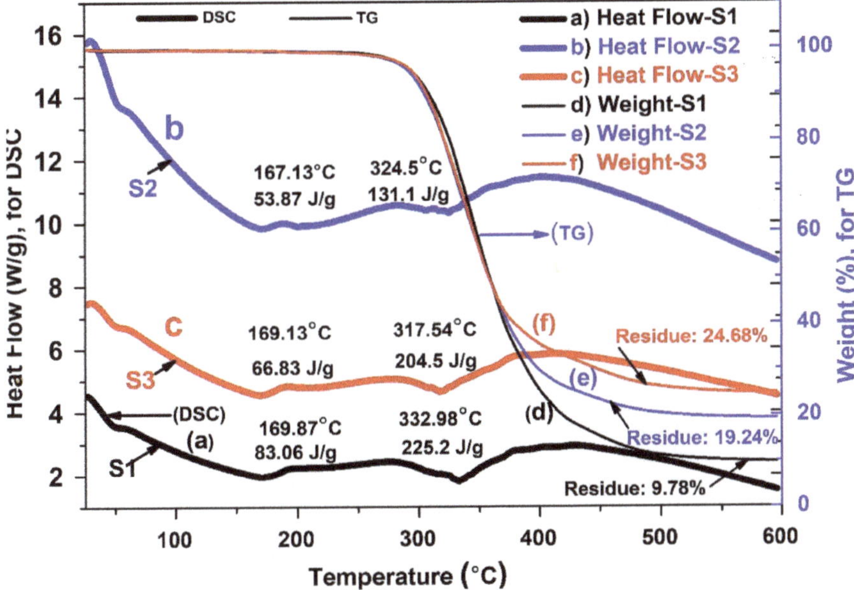

Figure 7. Simultaneous thermogravimetric analysis/differential scanning calorimetry (TGA/DSC) of the three experimental groups comprising plain: (**a**–**c**) heat flow and (**d**–**f**) residual weight of plain and composite layers coated on Ti substrates.

It has been widely reported that coatings can easily delaminate due to a weak adhesion strength, significantly affecting the performance and stability of such coated biomedical devices [49]. In medical device development, adhesion stability between the metal substrate and coating is, therefore, vital to the successful use of implant. The degree of adhesion between the polymer layer and Ti substrate depends intimately on the stability and subsequently longevity of coatings. In the present study, the crosscut (known as tape) adhesion test was conducted to investigate the samples' coating adhesion quality. Table 2 demonstrates the mean of the adhesion performance measurements. Interestingly, as expected, the S2 and S3 groups showed superior adhesion performance (5B, Table 2). It is assumed that the increased surface hydrophilicity (wettability) and the morphological changes caused by the alkali treatment improved the interfacial strength between the PU molecules and the treated Ti substrates.

Table 2. Cross-cult adhesion test results of Mg particles/PU and HAp-Mg particles/PU composite coatings formed on alkali-treated Ti (assessed by ASTM (D3359, 2010), a standard protocol).

Samples	Rating of Extent of Adhesion as per ASTM (D3359, 2010)	Percent Area Removed *	Surface Appearance after Adhesion Testing
S2	5B	0% (none)	
S3	5B	0% (none)	

* Percent area removed was determined based on the classification of adhesion test results of D3359 ASTM standard for measuring adhesion by tape test, according to the following criteria: 5B: The edges of the cuts are completely smooth; none of the squares of the lattice are detached.

The degree of surface roughness and charge (wettability) directly influence coating adhesion, with increased surface polarity and roughness resulting in increased reactivity between the substrate and coating. For example, we observed that very fine needle-like hairs were produced on the alkali-treated Ti (inset of Figure 1a). The high surface area of these features appeared to directly improve the adhesion of the deposited PU film [10]. The incorporation of HAp NPs and Mg microparticles into the PU film did not show negative effects on the adhesion performance of the deposited composite film, but instead also improved the adhesion strength.

It is well-known that the adhesion properties between a polymer coating and a metallic surface involves either physical or chemical bonding. In this case, chemical bonding involves a chemical interaction between the deposited PU molecules and species on the alkali-treated Ti surface. Indeed, the surface polarity of the Na_2TiO_3 gel layer formed on the Ti is higher than the untreated Ti surfaces, where the surface comprises an inert amorphous titanium dioxide film. The conversion of the Na_2TiO_3 gel layer to titanium-hydroxide (Ti-OH) after contact with an aqueous environment (such as water vapor) has been reported in our previous work [39], showing that this type of reaction is thermodynamically reasonable. The presence of OH on a treated Ti substrate may confer interesting properties to the Ti surface. For instance, the alkali-treated Ti surface may conduct as a base when in contact with a PU film. It is therefore suggested that the PU film, which has an acidic group, might strongly interact with the Na_2TiO_3 gel layer. These properties make this treatment a very attractive method to enhance the performance of polymer coatings. In fact, this kind of interaction may directly enhance the adhesion strength, as it has been illustrated that electrostatic force develops at the interface between materials that possess different electronic band structures [50]. Thus, the adhesion strength/bonding is strong in the case of the deposition of PU film on the active treated Ti surface.

This phenomenon can be discussed in light of the large number of free ends on the chain of PU, as it has been well discussed in our previous report as well [4], which provide a large number of free carboxyl groups for electrostatic interactions with the alkali-treated titanium surface, as shown in Figure 2. To support our hypothesis, Fourier-transform infrared spectroscopy (FTIR) analyses were used to investigate the molecular level interaction between the film coatings and substrates, as shown in Figure 8. A PO_4^{3-}

peak (566 cm^{-1}) was present in the S3 group, similar to the one present in HAp. As is the case with biological HAp, peaks for carbonate (1411 and 1136 cm^{-1}) were also present in the S3 group sample. The wide peak around 3361 cm^{-1} for the S1 and S2 groups were due to the presence of both hydrogen bonds and a free –NH functional group. After the addition of HAp, there was an increase in absorption due to the H-bonded –NH. The peak around 3500 cm^{-1} indicated the presence of –OH groups, which remained broad due to hydrogen bonding. The peak around 2927 cm^{-1} indicated –CH stretching. The sharp peak around 1136 cm^{-1} indicated the presence of the free –C=O group of S0, which reduced upon the addition of Mg (S1-Mg). When HAp was added, the peak again increased, indicating a shifting of interaction from Mg (with –C=O), to Mg with HAp. The peak around 1100 cm^{-1} indicated the presence of an ether linkage, while a peak for Mg(OH)$_2$ was also present at 510 cm^{-1} in both S2 and S3 group samples [51].

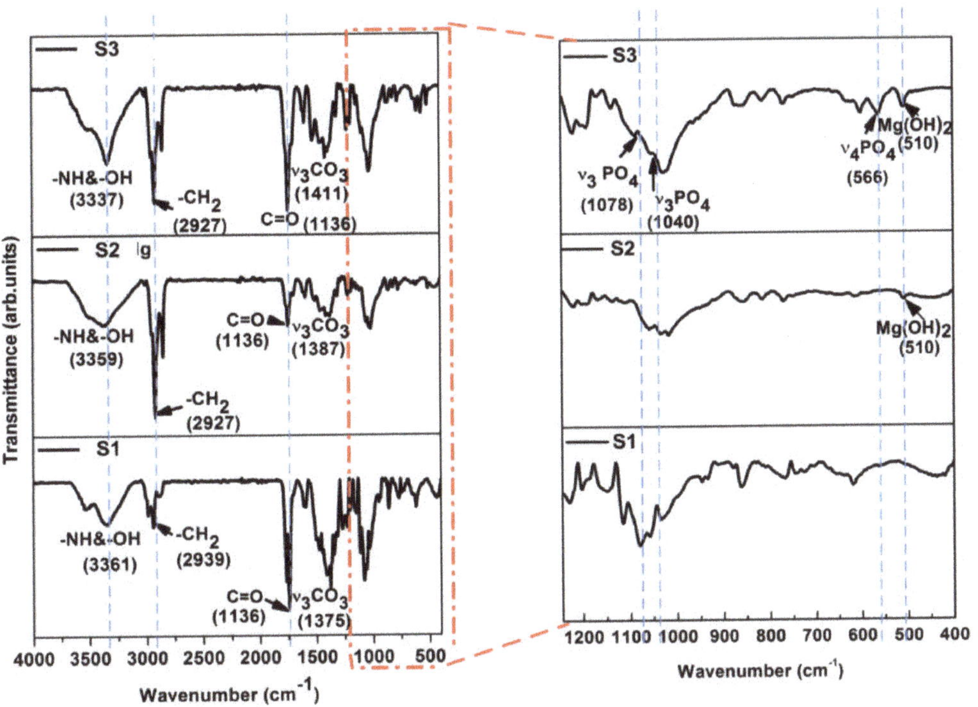

Figure 8. FT-IR spectra in transmission mode of a typical sample surface, comprising Ti coated with PU with and without Mg and HAp particles, after a 5 M NaOH chemical treatment at 60 °C for 24 h of the Ti samples.

The evaluation of cell/biomaterial interactions through the assessment of in vitro cytotoxicity test is one of the most fundamental initial tests for developed biomaterials [52]. Furthermore, the successful attachment of cells is an important part of this process and can strongly affect the subsequent cellular and tissue response. Figure 9 shows the cell morphology after three and five days of cell culture. The obtained images show the differential interaction of the cells with the different surfaces (some focusing issues arose from partial film detachment during the fixing, staining and dehydration processes).

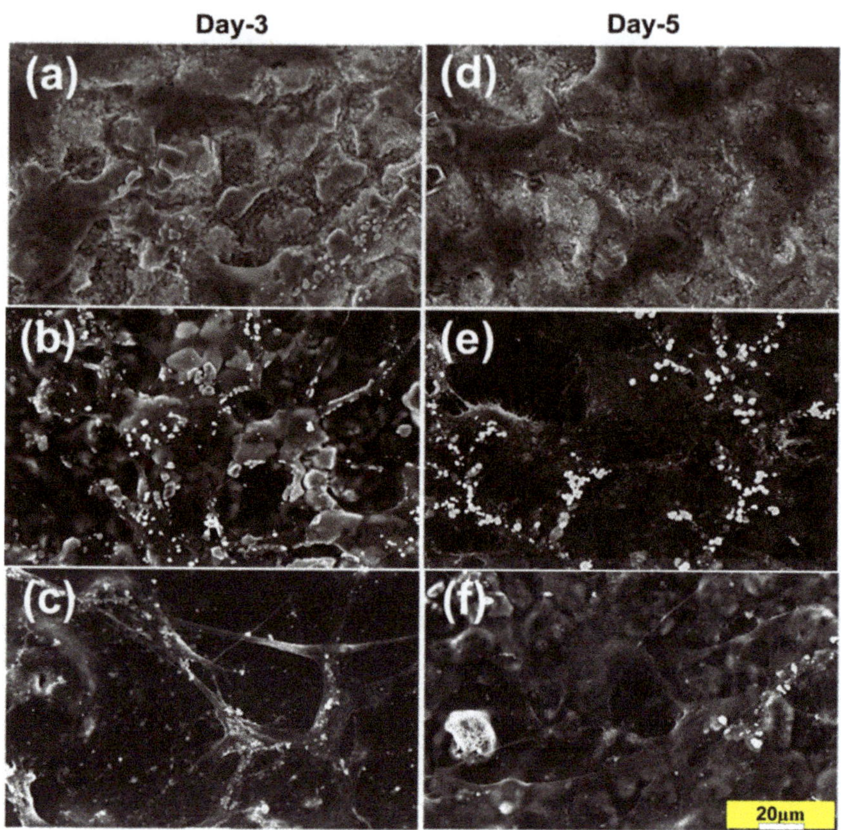

Figure 9. SEM images of MC3T3 osteoblastic cells lines cultured for three and five days on the treated and treated Ti surfaces of S0 (**a,d**), S2 (**b,e**) and S3 group (**c,f**). Scale bar 20 μm.

For the S0 group, only a few, after three days of culture and after five days poorly spread cells were observed, no additional changes in cell morphology were noticed (Figure 9). The coating remained intact throughout the culture period. In the S2 and S3 groups, there were similarly no defects, detachments, or other degradations in the coating after 5 days of culture (Figure 9b–f), and a clear cellular spreading was observed, particularly in group S3. These results suggest that the S3 group had the most cytocompatible characteristics among the tested groups (Figure 9e,f). The S0 group possessed a higher surface roughness (1.110 ± 0.015 μm) than the S2 group (0.611 ± 0.020 μm), which would typically indicate that it should have superior performance in terms of cell migration, spreading. However, the S2 group surface also possessed a lower stiffness than the S1 surface, and indeed previous studies [53] showed that cell migration and focal adhesion are regulated by substrate flexibility. A study by Pelham and colleagues [54] found that cell spreading/migration could be guided by surface topography and physical interactions between the cell and the materials' substrate. Their study concluded that changes in tissue rigidity and flexibility could play an important role in controlling cell spreading and migration. Additionally, cell spreading, and focal adhesions were affected by a change in mechanical properties of the extracellular matrix (ECM), which influenced cytoskeletal stiffness in vitro [52]. Ti-based implants possess a high stiffness that is responsible for the commonly problematic stress shielding phenomenon (i.e., there is a poor integration of the Ti implant with surrounding host tissues). Overall, it appears that the PU coating of alkali-treated Ti surfaces may regulate the adhesion and interactions of osteoblastic cells and the substrate.

Results of the MTT-assay confirmed that cell proliferation occurred over the 3 and 5 days of cell culture (Figure 10a). On day 3, it was observed that the S0 samples showed less cell proliferation than the S2 and S3 samples. After 5 d of culture, all groups showed a significant increase, and maintained over 80% viability. It has previously been reported that the formation of sodium titanate on the surface of Ti implants that are implanted in the body may have a harmful effect on cellular response because of the excessive release of sodium ions and the formation of narrow pore spaces [55]. We also observed that the Na^+ ions released from the alkali-treated Ti surfaces had a negative effect on cell proliferation, despite the surface having an increased hydrophilicity [2]. However, the biocompatibility of the material was improved after coating, particularly with the S3 group samples, which performed the best of all of the groups.

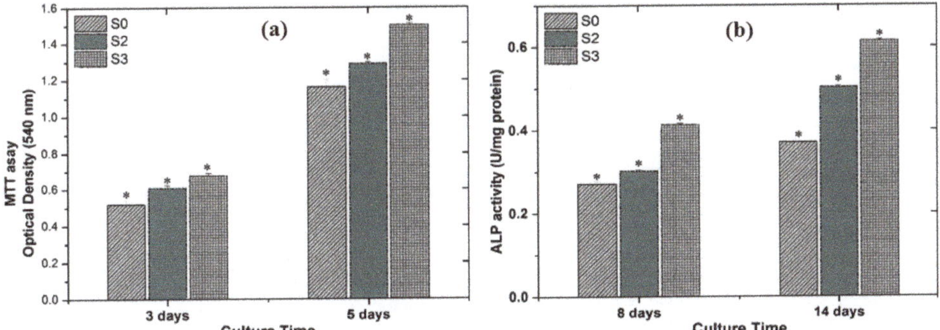

Figure 10. (a) Cell proliferation assay after 3 and 5 d of culture and (b) Alkaline phosphate activity (ALP) (U/mg protein) after 8 and 14 d of culture; cells were seeded on coated alkaline treated Ti samples (n = 3, * indicates $p \leq 0.05$).

Alkaline phosphate (ALP) is an early marker of osteoblast differentiation. We observed a clear increase in ALP activity after 8 and 14 d of culture, with ALP activity showing a significant increase in group S3 by day 14 (Figure 10b). Overall, the best outcomes in terms of cell proliferation and differentiation were observed in group S3. This observation indicates that there is a positive correlation between Mg^{2+} ion release into the media and an increase in osteogenic response.

The effect of Mg^{2+} ions on bone growth and bone formation was recently investigated [56]. The studies of the role of Mg^{2+} ions in accelerating bone formation have showed that the ions not only enhance bone adhesion and bone healing but also aid in the regulation and acceleration formation of bone marrow cells through the enhancement of BMP-receptor recognition, Smad signaling pathways, and/or the upregulation of neuronal calcitonin gene-related polypeptide (CGRP) [57,58]. The stimulation of bone formation by Mg^{2+} ions has been widely reported including ours, primarily through enhancing the activity of osteoblast, including adhesion/attachment, growth, and differentiation of these cells [30]. Mg^{2+} ions are actively involved in the process minerals formation to control bone formation and resorption [23]. Accordingly, the hypothesis that the S3 samples would increase MC3T3-E1 proliferation and differentiation compared with S0 samples, is accepted. Finally, it appears that coatings of both plain PU and PU doped with metallic Mg particles in a thin film deposited on a Ti implant are safe and effective in an in vitro cell culture model. Using a HAp/Mg/PU film coating on the Ti implants also further increased the level of osteoblast cell proliferation and differentiation, benefiting from the synergistic effects of both Mg and HAp. We have planned future work to investigate the microenvironment of the coated implants more comprehensively, and furthermore plan to conduct in vivo studies where we evaluate the coating durability, and the osteogenesis, osteoinduction, and osseointegration of these implants.

4. Conclusions

In this study, alkali-treated Ti substrates were coated with thin films of PU, PU with Mg particles, and PU with Mg and hydroxyapatite (HAp) particles using a dip-coating technology. The coatings were stable and did not delaminate from the implant surfaces. The use of dip-coating technology allowed us to create novel composites, introducing biologically compatible metal particles (Mg) and HAp onto alkali-treated Ti in a short time using a facile process. The morphology of the modified Ti surface interacted very favorably with the HAp NPs, increasing the adhesive strength between the substrate and the coating. Additionally, the surface charge (wettability) of the HAp NPs appeared to improve the interfacial strength by interacting with the sodium titanate gel layer formed on the treated Ti substrates. The HAp/Mg/PU group achieved the most positive effects in terms of enhancing preosteoblast adhesion, proliferation, and differentiation, which can be attributed to the bioactivity and biocompatibility of both the HAp particles and Mg^{2+} ions released from the metal particles. Overall, our results suggest that HAp/Mg/PU coated alkali-treated Ti is highly suitable as a biodegradable and bioactive orthopedic implant material.

Supplementary Materials: The following are available online at https://www.mdpi.com/article/10.3390/nano11051129/s1, Sample preparation, Wet chemical synthesis of HAp, Coating process, Characterization, Adhesion performance, Osteoblast cell response.

Author Contributions: Conceptualization, A.A.-h. and M.A.; methodology, M.A.; software, M.A.; validation, M.A.; original draft preparation, M.A. and A.A.-h.; writing—review and editing, M.K.H., M.B., and S.I.; visualization, S.I.; supervision, A.A.-h. and S.I. All authors have read and agreed to the published version of the manuscript.

Funding: This research received no external funding.

Institutional Review Board Statement: Not applicable.

Informed Consent Statement: Not applicable.

Acknowledgments: The authors acknowledge the facilities, and the scientific and technical assistance, of the Australian Microscopy & Microanalysis Research Facility at the Centre for Microscopy and Microanalysis, The University of Queensland.

Conflicts of Interest: The authors declare no conflict of interest.

References

1. Park, K.-D.; Lee, B.-A.; Piao, X.-H.; Lee, K.-K.; Park, S.-W.; Oh, H.-K.; Kim, Y.-J.; Park, H.-J. Effect of magnesium and calcium phosphate coatings on osteoblastic responses to the titanium surface. *J. Adv. Prosthodont.* **2013**, *5*, 402–408. [CrossRef]
2. Young Kim, S.; kyoung Kim, Y.; song Park, I.; chun Jin, G.; sung Bae, T.; ho Lee, M. Effect of alkali and heat treatments for bioactivity of TiO_2 nanotubes. *Appl. Surf. Sci.* **2014**, *321*, 412–419.
3. Roy, P.; Berger, S.; Schmuki, P. TiO_2 nanotubes: Synthesis and applications. *Angew. Chem. Int. Ed.* **2011**, *50*, 2904–2939. [CrossRef]
4. Abdal-hay, A.; Agour, M.; Kim, Y.-K.; Lee, M.-H.; Hassan, M.K.; El-Ainin, H.A.; Hamdy, A.S.; Ivanovski, S. Magnesium-particle/polyurethane composite layer coating on titanium surfaces for orthopedic applications. *Eur. Polym. J.* **2019**, *112*, 555–568. [CrossRef]
5. Lim, Y.J.; Oshida, Y.; Andres, C.J.; Barco, M.T. Surface characterizations of variously treated titanium materials. *Int. J. Oral Maxillofac. Implants* **2001**, *16*, 333–342. [PubMed]
6. Lopes, M.A.; Monteiro, F.; Santos, J.; Serro, A.; Saramago, B. Hydrophobicity, surface tension, and zeta potential measurements of glass—Reinforced hydroxyapatite composites. *J. Biomed. Mater. Res.* **1999**, *45*, 370–375. [CrossRef]
7. Webster, T.J.; Ergun, C.; Doremus, R.H.; Siegel, R.W.; Bizios, R. Specific proteins mediate enhanced osteoblast adhesion on nanophase ceramics. *J. Biomed. Mater. Res.* **2000**, *51*, 475–483. [CrossRef]
8. Ferraz, M.; Monteiro, F.; Serro, A.; Saramago, B.; Gibson, I.R.; Santos, J. Effect of chemical composition on hydrophobicity and zeta potential of plasma sprayed $HA/CaO-P_2O_5$ glass coatings. *Biomaterials* **2001**, *22*, 3105–3112. [CrossRef]
9. Botelho, C.; Lopes, M.; Gibson, I.R.; Best, S.; Santos, J. Structural analysis of Si-substituted hydroxyapatite: Zeta potential and X-ray photoelectron spectroscopy. *J. Mater. Sci. Mater. Med.* **2002**, *13*, 1123–1127. [CrossRef] [PubMed]

0.   Abdal-Hay, A.; Hamdy, A.S.; Khalil, K.A.; Lim, J.H. A novel simple one-step air jet spinning approach for deposition of poly (vinyl acetate)/hydroxyapatite composite nanofibers on Ti implants. *Mater. Sci. Eng. C* **2015**, *49*, 681–690. [CrossRef]

1.   Vlacic-Zischke, J.; Hamlet, S.; Friis, T.; Tonetti, M.; Ivanovski, S. The influence of surface microroughness and hydrophilicity of titanium on the up-regulation of TGFβ/BMP signalling in osteoblasts. *Biomaterials* **2011**, *32*, 665–671. [CrossRef]

2.   Abdal-hay, A.; Abbasi, N.; Gwiazda, M.; Hamlet, S.; Ivanovski, S. Novel polycaprolactone/hydroxyapatite nanocomposite fibrous scaffolds by direct melt-electrospinning writing. *Eur. Polym. J.* **2018**, *105*, 257–264. [CrossRef]

3.   Abdal-hay, A.; Fouad, H.; ALshammari, B.A.; Khalil, K.A. Biosynthesis of bonelike apatite 2D nanoplate structures using fenugreek seed extract. *Nanomaterials* **2020**, *10*, 919. [CrossRef]

4.   Porter, A.E.; Taak, P.; Hobbs, L.W.; Coathup, M.J.; Blunn, G.W.; Spector, M. Bone bonding to hydroxyapatite and titanium surfaces on femoral stems retrieved from human subjects at autopsy. *Biomaterials* **2004**, *25*, 5199–5208. [CrossRef] [PubMed]

5.   Khalil, K.A.; Kim, S.W.; Dharmaraj, N.; Kim, K.W.; Kim, H.Y. Novel mechanism to improve toughness of the hydroxyapatite bioceramics using high-frequency induction heat sintering. *J. Mater. Process. Technol.* **2007**, *187*, 417–420. [CrossRef]

6.   Barakat, N.A.; Khalil, K.; Sheikh, F.A.; Omran, A.; Gaihre, B.; Khil, S.M.; Kim, H.Y. Physiochemical characterizations of hydroxyapatite extracted from bovine bones by three different methods: Extraction of biologically desirable HAp. *Mater. Sci. Eng. C* **2008**, *28*, 1381–1387. [CrossRef]

7.   Lin, D.-Y.; Wang, X.-X. Preparation of hydroxyapatite coating on smooth implant surface by electrodeposition. *Ceram. Int.* **2011**, *37*, 403–406. [CrossRef]

8.   Szymura-Oleksiak, J.; Ślósarczyk, A.; Cios, A.; Mycek, B.; Paszkiewicz, Z.; Szklarczyk, S.; Stankiewicz, D. The kinetics of pentoxifylline release in vivo from drug-loaded hydroxyapatite implants. *Ceram. Int.* **2001**, *27*, 767–772. [CrossRef]

9.   Chen, B.; Zhang, T.; Zhang, J.; Lin, Q.; Jiang, D. Microstructure and mechanical properties of hydroxyapatite obtained by gel-casting process. *Ceram. Int.* **2008**, *34*, 359–364. [CrossRef]

20.   Abdal-hay, A.; Barakat, N.A.; Lim, J.K. Hydroxyapatite-doped poly (lactic acid) porous film coating for enhanced bioactivity and corrosion behavior of AZ31 Mg alloy for orthopedic applications. *Ceram. Int.* **2013**, *39*, 183–195. [CrossRef]

21.   Wang, C.; Wang, M. Electrochemical impedance spectroscopy study of the nucleation and growth of apatite on chemically treated pure titanium. *Mater. Lett.* **2002**, *54*, 30–36. [CrossRef]

22.   Xu, L.; Pan, F.; Yu, G.; Yang, L.; Zhang, E.; Yang, K. In vitro and in vivo evaluation of the surface bioactivity of a calcium phosphate coated magnesium alloy. *Biomaterials* **2009**, *30*, 1512–1523. [CrossRef] [PubMed]

23.   Ibasco, S.; Tamimi, F.; Meszaros, R.; Le Nihouannen, D.; Vengallatore, S.; Harvey, E.; Barralet, J.E. Magnesium-sputtered titanium for the formation of bioactive coatings. *Acta Biomater.* **2009**, *5*, 2338–2347. [CrossRef] [PubMed]

24.   Huynh, V.; Ngo, N.K.; Golden, T.D. Surface activation and pretreatments for biocompatible metals and alloys used in biomedical applications. *Int. J. Biomater.* **2019**, *2019*, 3806504. [CrossRef] [PubMed]

25.   Xuhui, Z.; Lingfang, Y.; Yu, Z.; Xiong, J. Hydroxyapatite coatings on titanium prepared by electrodeposition in a modified simulated body fluid. *Chin. J. Chem. Eng.* **2009**, *17*, 667–671.

26.   Rath, P.C.; Besra, L.; Singh, B.P.; Bhattacharjee, S. Titania/hydroxyapatite bi-layer coating on Ti metal by electrophoretic deposition: Characterization and corrosion studies. *Ceram. Int.* **2012**, *38*, 3209–3216. [CrossRef]

27.   Cifuentes, S.C.; Gavilán, R.; Lieblich, M.; Benavente, R.; González-Carrasco, J.L. In vitro degradation of biodegradable polylactic acid/magnesium composites: Relevance of Mg particle shape. *Acta Biomater.* **2016**, *32*, 348–357. [CrossRef]

28.   Shuai, C.; Li, Y.; Feng, P.; Guo, W.; Yang, W.; Peng, S. Positive feedback effects of Mg on the hydrolysis of poly-l-lactic acid (PLLA): Promoted degradation of PLLA scaffolds. *Polym. Test.* **2018**, *68*, 27–33. [CrossRef]

29.   Zimmermann, T.; Ferrandez-Montero, A.; Lieblich, M.; Ferrari, B.; González-Carrasco, J.L.; Müller, W.-D.; Schwitalla, A.D. In vitro degradation of a biodegradable polylactic acid/magnesium composite as potential bone augmentation material in the presence of titanium and PEEK dental implants. *Dent. Mater.* **2018**, *34*, 1492–1500. [CrossRef] [PubMed]

30.   Abdal-hay, A.; Dewidar, M.; Lim, J.; Lim, J.K. Enhanced biocorrosion resistance of surface modified magnesium alloys using inorganic/organic composite layer for biomedical applications. *Ceram. Int.* **2014**, *40*, 2237–2247. [CrossRef]

31.   Barrere, F.; Van Der Valk, C.; Meijer, G.; Dalmeijer, R.; De Groot, K.; Layrolle, P. Osteointegration of biomimetic apatite coating applied onto dense and porous metal implants in femurs of goats. *J. Biomed. Mater. Res. Part B Appl. Biomater.* **2003**, *67*, 655–665. [CrossRef]

32.   Guelcher, S.A. Biodegradable polyurethanes: Synthesis and applications in regenerative medicine. *Tissue Eng. Part. B Rev.* **2008**, *14*, 3–17. [CrossRef] [PubMed]

33.   Wang, P.; Li, C.; Gong, H.; Jiang, X.; Wang, H.; Li, K. Effects of synthesis conditions on the morphology of hydroxyapatite nanoparticles produced by wet chemical process. *Powder Technol.* **2010**, *2032*, 315–321. [CrossRef]

34.   Song, G.; Song, S. A possible biodegradable magnesium implant material. *Adv. Eng. Mater.* **2007**, *9*, 298–302. [CrossRef]

35.   Kokubo, T.; Kim, H.-M.; Kawashita, M. Novel bioactive materials with different mechanical properties. *Biomaterials* **2003**, *24*, 2161–2175. [CrossRef]

36.   Lu, X.; Wang, Y.-B.; Liu, Y.-R.; Wang, J.-X.; Qu, S.-X.; Feng, B.; Weng, J. Preparation of HA/chitosan composite coatings on alkali-treated titanium surfaces through sol-gel techniques. *Mater. Lett.* **2007**, *61*, 3970–3973. [CrossRef]

37.   Cui, W.; Jin, L.; Zhou, L. Surface characteristics and electrochemical corrosion behavior of a pre-anodized microarc oxidation coating on titanium alloy. *Mater. Sci. Eng. C* **2013**, *33*, 3775–3779. [CrossRef]

38. Xie, Y.; Ao, H.; Xin, S.; Zheng, X.; Ding, C. Enhanced cellular responses to titanium coating with hierarchical hybrid structure *Mater. Sci. Eng. C* **2014**, *38*, 272–277. [CrossRef]

39. Abdal-hay, A.; Gulati, K.; Fernandez-Medina, T.; Qian, M.; Ivanovski, S. In situ hydrothermal transformation of titanium surface into lithium-doped continuous nanowire network towards augmented bioactivity. *Appl. Surf. Sci.* **2020**, *505*, 144604. [CrossRef]

40. Abdal-hay, A.; Amna, T.; Lim, J.K. Biocorrosion and osteoconductivity of PCL/nHAp composite porous film-based coating of magnesium alloy. *Solid State Sci.* **2012**, *18*, 131–140. [CrossRef]

41. Da Conceicao, T.; Scharnagl, N.; Blawert, C.; Dietzel, W.; Kainer, K. Surface modification of magnesium alloy AZ31 by hydrofluoric acid treatment and its effect on the corrosion behaviour. *Thin Solid Film.* **2010**, *518*, 5209–5218. [CrossRef]

42. da Conceição, T.F. Corrosion Protection of Magnesium AZ31 Alloy Sheets by Polymer Coatings. Ph.D. Thesis, Technische Universität, Berlin, Germany, 2011.

43. Siau, S.; Vervaet, A.; Degrande, S.; Schacht, E.; Van Calster, A. Dip coating of dielectric and solder mask epoxy polymer layers for build-up purposes. *Appl. Surf. Sci.* **2005**, *245*, 353–368. [CrossRef]

44. Fang, H.-W.; Li, K.-Y.; Su, T.-L.; Yang, T.C.-K.; Chang, J.-S.; Lin, P.-L.; Chang, W.-C. Dip coating assisted polylactic acid deposition on steel surface: Film thickness affected by drag force and gravity. *Mater. Lett.* **2008**, *62*, 3739–3741. [CrossRef]

45. Yimsiri, P.; Mackley, M.R. Spin and dip coating of light-emitting polymer solutions: Matching experiment with modelling. *Chem. Eng. Sci.* **2006**, *61*, 3496–3505. [CrossRef]

46. Shukla, A.; Balasubramaniam, R. Effect of surface treatment on electrochemical behavior of CP Ti, Ti–6Al–4V and Ti–13Nb–13Zr alloys in simulated human body fluid. *Corros. Sci.* **2006**, *48*, 1696–1720. [CrossRef]

47. Xue, W.; Liu, X.; Zheng, X.; Ding, C. In vivo evaluation of plasma-sprayed titanium coating after alkali modification. *Biomaterials* **2005**, *26*, 3029–3037. [CrossRef] [PubMed]

48. Abdal-hay, A.; Bartnikowski, M.; Hamlet, S.; Ivanovski, S. Electrospun biphasic tubular scaffold with enhanced mechanical properties for vascular tissue engineering. *Mater. Sci. Eng. C* **2018**, *82*, 10–18. [CrossRef]

49. Burke, M.; Clarke, B.; Rochev, Y.; Gorelov, A.; Carroll, W. Estimation of the strength of adhesion between a thermoresponsive polymer coating and nitinol wire. *J. Mater. Sci. Mater. Med.* **2008**, *19*, 1971–1979. [CrossRef] [PubMed]

50. Catauro, M.; Bollino, F.; Papale, F.; Giovanardi, R.; Veronesi, P. Corrosion behavior and mechanical properties of bioactive sol-gel coatings on titanium implants. *Mater. Sci. Eng. C* **2014**, *43*, 375–382. [CrossRef]

51. Feng, J.; Chen, Y.; Liu, X.; Liu, T.; Zou, L.; Wang, Y.; Ren, Y.; Fan, Z.; Lv, Y.; Zhang, M. In-situ hydrothermal crystallization $Mg(OH)_2$ films on magnesium alloy AZ91 and their corrosion resistance properties. *Mater. Chem. Phys.* **2013**, *143*, 322–329. [CrossRef]

52. Halliday, N.L.; Tomasek, J.J. Mechanical properties of the extracellular matrix influence fibronectin fibril assembly in vitro. *Exp. Cell Res.* **1995**, *217*, 109–117. [CrossRef]

53. Pelham, R.J.; Wang, Y.-L. Cell locomotion and focal adhesions are regulated by substrate flexibility. *Proc. Natl. Acad. Sci. USA* **1997**, *94*, 13661–13665. [CrossRef]

54. Lo, C.-M.; Wang, H.-B.; Dembo, M.; Wang, Y.-L. Cell movement is guided by the rigidity of the substrate. *Biophys. J.* **2000**, *79*, 144–152. [CrossRef]

55. Pattanayak, D.K.; Yamaguchi, S.; Matsushita, T.; Kokubo, T. Nanostructured positively charged bioactive TiO_2 layer formed on Ti metal by NaOH, acid and heat treatments. *J. Mater. Sci. Mater. Med.* **2011**, *22*, 1803–1812. [CrossRef]

56. Zhang, Y.; Xu, J.; Ruan, Y.C.; Yu, M.K.; O'Laughlin, M.; Wise, H.; Chen, D.; Tian, L.; Shi, D.; Wang, J. Implant-derived magnesium induces local neuronal production of CGRP to improve bone-fracture healing in rats. *Nat. Med.* **2016**, *22*, 1160. [CrossRef]

57. Brown, A.; Zaky, S.; Ray, H.; Sfeir, C. Porous magnesium/PLGA composite scaffolds for enhanced bone regeneration following tooth extraction. *Acta Biomater.* **2015**, *11*, 543–553. [CrossRef] [PubMed]

58. Alvarez-Lopez, M.; Pereda, M.D.; del Valle, J.A.; Fernandez-Lorenzo, M.; Garcia-Alonso, M.C.; Ruano, O.A.; Escudero, M.L. Corrosion behaviour of AZ31 magnesium alloy with different grain sizes in simulated biological fluids. *Acta Biomater.* **2010**, *6*, 1763–1771. [CrossRef] [PubMed]

Communication

TiB Nanowhisker Reinforced Titanium Matrix Composite with Improved Hardness for Biomedical Applications

Joseph A. Otte [1,*], Jin Zou [1,2], Rushabh Patel [1], Mingyuan Lu [1] and Matthew S. Dargusch [1,*]

[1]   School of Mechanical and Mining Engineering, The University of Queensland, Brisbane 4072, Australia; j.zou@uq.edu.au (J.Z.); r.patel@uq.edu.au (R.P.); m.lu1@uq.edu.au (M.L.)

[2]   Centre for Microscopy and Microanalysis, The University of Queensland, Brisbane 4072, Australia

*   Correspondence: joseph.otte@uqconnect.edu.au (J.A.O.); m.dargusch@uq.edu.au (M.S.D.); Tel.: +61-4-359-14697 (J.A.O.); +61-7-334-69225 (M.S.D.)

Received: 4 November 2020; Accepted: 8 December 2020; Published: 10 December 2020

Abstract: Titanium and its alloys have been employed in the biomedical industry as implants and show promise for more broad applications because of their excellent mechanical properties and low density. However, high cost, poor wear properties, low hardness and associated side effects caused by leaching of alloy elements in some titanium alloys has been the bottleneck to their wide application. TiB reinforcement has shown promise as both a surface coating for Ti implants and also as a composite reinforcement phase. In this study, a low-cost TiB-reinforced alpha titanium matrix composite (TMC) is developed. The composite microstructure includes ultrahigh aspect ratio TiB nanowhiskers with a length up to 23 μm and aspect ratio of 400 and a low average Ti grain size. TiB nanowhiskers are formed in situ by the reaction between Ti and BN nanopowder. The TMC exhibited hardness of above 10.4 GPa, elastic modulus above 165 GPa and hardness to Young's modulus ratio of 0.062 representing 304%, 170% and 180% increases in hardness, modulus and hardness to modulus ratio, respectively, when compared to commercially pure titanium. The TiB nanowhisker-reinforced TMC has good biocompatibility and shows excellent mechanical properties for biomedical implant applications.

Keywords: nano-composite; microstructure; nanoindentation; bone implants; powder metallurgy

---

1. Introduction

As our population ages the need for effective bone implants as treatments for arthritis and other joint-related pain is of growing importance [1,2]. Current projections indicate that in the United States replacement surgery for hips and knee arthroplasties will increase by over 600% to more than 4 million operations by 2030 [3]. Many of these surgeries arise due to failure of the original implant and need for a second replacement surgery which can lead to several associated health concerns. Pure titanium (CP-Ti) stands out as a good material for load-bearing joint replacements because of its strong specific properties, corrosion resistance and good biocompatibility [4,5]. However, poor wear resistance of Ti limits its usefulness as the removal of the surface of the implant over time not only contributes to the breakdown of the implant but can also lead to inflammation and related immune responses in the affected area [6]. Attempts to improve strength and wear resistance of Ti through alloying has introduced elements which are prone to leach out over time into the body, limiting their safe use and lifetime [7]. A Ti-based material that retains the aforementioned strengths in biocompatibility and corrosion performance with increased hardness and wear resistance will be a strong candidate as an implant material.

One possible avenue for improving the wear properties is by the surface treatment of Ti, typically with a coating of hydroxyapatite, by methods such as dip coating, plasma spraying and electrochemical deposition [8–11] amongst others. Hydroxyapatite is a constituent of hard tissues and so interfaces well with the human body when implanted [12]. However, these coatings add an additional cost to the already expensive manufacture of Ti implants and in addition have low bond strength to Ti and are prone to peel off during long-term service [4]. With these facts, producing an uncoated Ti component with strong wear properties is desirable.

Titanium matrix composites (TMCs) have many of the desirable properties for bone implants, with significantly improved hardness and wear properties when compared to CP-Ti. By reinforcing with stiff and strong ceramic particles or fibres, Ti has shown significantly improved strength, hardness and wear resistance [13–19]. In addition, many of the most promising reinforcements (Ti_xB, TiC and Ti_xN) can be manufactured 'in situ' meaning that the reinforcement phase is formed during processing (typically facilitated by a reaction at high temperature), which reduces overall process steps and leads to strong bonding at the reinforcement interface [20–22]. Among these materials TiB stands out, as it naturally forms a hexagonal whisker morphology which offers an improved strengthening effect when compared to particle reinforcement [23]. TiB is also thermodynamically stable for room temperature operations, leading to stable overall mechanical properties.

There are important concerns about the safety and biocompatibility of whisker reinforced composites for implant materials. Several studies have sought to investigate the cytotoxicity and osteointegration of cells on TiB composites. A 29 vol% TiB whisker reinforced CP-Ti TMC has been tested for biocompatibility and has shown, amongst other results, to have a H index similar to CP-Ti and approximately a third of Ti-6Al-4V indicating that the TiB had no adverse effect on biocompatibility and is not cytotoxic [24]. Similarly, a TiB-TiN Ti6Al4V coating has been shown to have strong biocompatibility and good osteointegration with 15 wt% BN addition [25]. Further, towards dental implant applications, human gingival fibroblasts and osteoblasts on a CP-Ti sample coated with a TiB whisker composite showed similar performance to CP-Ti only indicating their suitability as a biomaterial with up to 10 wt% B addition [26]. Subsequently, studies on the corrosion behaviour of TiB whisker reinforced composites have shown that TiB composites exhibit stronger passive film formation and reduced corrosion when compared to CP-Ti in simulated body fluids (Hank's solution 37 °C) [27]. This suggests that TiB reinforced composites will remain stable without significant corrosion, preventing leaching of harmful foreign elements. There is very little chance that TiB nanowhiskers will be released into the body. The strong bond between the Ti matrix and TiB produced by an in situ reaction has been shown to significantly reduce fibre pullout and improves the safety of these materials [28]. Further study is required in this area and 'in vivo' studies are recommended to identify any potential hazards of TiB nanowhisker-reinforced composites, particularly when subjected to long periods of articulating wear. However, these early studies show that TiB nanowhisker-reinforced TMCs have strong potential to be a safe, viable and long-lasting biomaterial for implant applications.

CP-Ti samples have found strong industrial application as biomedical implants particularly in low wear areas like the femoral stem [28]. However, there are two key problems that need to be addressed. Firstly, a high hardness and wear resistant material is required for high wear and articulating applications such as the femoral head [29]. Secondly, the complexity of coatings coupled with the risk of poor bonding at the surface increases the cost and can reduce the lifetime of coated implants, leading to expensive and dangerous replacement surgery [4]. TiB reinforced matrix composites are in a position to overcome these challenges as a single material can fill the role of a wide range of load bearing implants and with relatively low cost and long lifetime. However, there is a need to develop a thorough understanding of the manufacture of these composites and the optimization of the in situ reaction before they can find useful industrial application in the biomedical industry.

In this study, we demonstrate a cost-effective method to manufacture high aspect ratio TiB nanowhisker reinforced TMCs for load bearing biomedical applications, with mechanical properties investigated by nanoindentation.

2. Materials and Methods

In this study, TMC samples were prepared by a process of powder mixing followed by cold compaction and vacuum sintering. CP-Ti gas atomized spherical powders (0–45 µm) were combined with BN hexagonal nanopowders (65–75 nm). One batch of samples was produced with 5 vol% BN and another with 10 vol% BN. Powders were then mixed, compacted and vacuum sintered using the same method as previously reported [15]. Samples were sintered at peak temperatures ranging from 1050 to 1200 °C with the holding times ranging from 2 to 6 h to facilitate the in situ reaction between Ti and BN for the formation of TiB. The furnace heating and cooling rate was fixed at 2 °C/min for all samples. In addition, to reduce the risk of chloride impurities and oxidation, samples were sintered on a bed of yttrium oxide and surrounded with blocks of sponge Ti.

The microstructural characteristics of the manufactured samples were analysed using scanning electron microscopy (SEM, Hitachi SU3500, Brisbane, Australia) and transmission electron microscopy (TEM, Hitachi HF5000, Brisbane, Australia). To prepare samples for SEM imaging, conventional metallographic polishing techniques were used with some SEM specimens etched in a modified Kroll's solution (86% H_2O, 10% HNO_3 and 4% HF) for 30 s. TEM specimens were prepared by two methods. Firstly, bulk samples after sintering were deeply etched with the modified Kroll's solution for 3 min before being ultrasonicated in ethanol to extract and disperse nanowhiskers in solution. This solution was then dripped onto standard carbon grids, so isolated nanowhiskers could be investigated and measured. Secondly, TEM lamellae were prepared from bulk sintered samples with a FEI Scios Dual Beam FIB using a standard Ga ion beam cut and lift out technique [30]. Sample phases were characterized using a Rigaku SmartLab X-Ray diffractometer (XRD) with a 9 kW rotating Cu anode source. 45 kV and 200 mA were used to generate a 300 µm beam with data recorded by continuous scan from 20–80° 2θ angle in 0.02° increments.

Mechanical properties of the manufactured samples were investigated by nanoindentation tests using a Hysitron Triboindenter with a Berkovich probe. Nanoindentation was chosen in favour of tensile tests because local property measurement facilitates the study of TiB size and microstructural effects, without being affected by other sample variables present in sintered samples, such as porosity. A 12 mN load was chosen as tests conducted with this load in a previous work gave mechanical properties on CP-Ti comparable to that obtained from tensile tests [15]. Oliver–Pharr (OP) analysis was used to determine hardness (H) and reduced elastic modulus (E_r) from the load (P)—depth (h) curve. Elastic modulus (E) was then calculated using Equation (1) as shown below, where v_i and E_i are the Poisson's ratio and elastic modulus of the diamond indenter (0.07 and 1140 GPa) and v and E are the samples Poisson's ratio and elastic modulus. Details of OP analysis method can be found in [31]. Samples were tested with at least 100 indents in a 10 × 10 matrix with 10 µm spacing to ensure the result was not affected by the plastic zone of the adjacent indents.

$$\frac{1}{E_r} = \frac{1 - v^2}{E} + \frac{1 - v_i^2}{E_i} \tag{1}$$

3. Results and Discussion

3.1. Powder Mixing

Powder mixing plays an important role in the manufacture of composite materials [32]. Firstly, proper mixing ensures a uniform distribution of the reinforcement phase in the product part, leading to more uniform mechanical properties. Secondly, when considering in situ processes, well-mixed powders have a higher and more consistent contact area between the two powders which will undergo the in situ reaction. Strong interfacial contact between the reacting particles reduces the risk of any particles remaining unreacted after processing and increases the nucleation rate, acting to refine the microstructure [33]. Figure 1 shows the powders before and after mixing, in which Figure 1a shows the BN nanoparticles as purchased from Nanografi. The BN particles have a plate-like morphology, up to 100 nm in diameter and with a typical thickness of <10 nm. Figure 1b is a low magnification SEM image showing the spherical CP-Ti powders used in this work. Figure 1c,d are secondary electron (SE) and back-scatter electron (BSE) images, respectively, showing the powders after mixing for 4 h in a low energy Turbula shaker mixer with a 3:1 ratio of steel balls [15]. Figure 1c shows the fine BN powders are well distributed across the surface of the CP-Ti and similarly in Figure 1d, where dark patches are indicative of BN. It is clear from this image that BN has been distributed around the surface of CP-Ti particles. Figure 1c also shows some BN plates embedded into the Ti particle which will further promote their in situ reaction. In addition, it is clear that BN nanoparticle agglomerates like those shown in Figure 1a were largely broken up during mixing. Therefore, this mixing process has been effective to distribute BN.

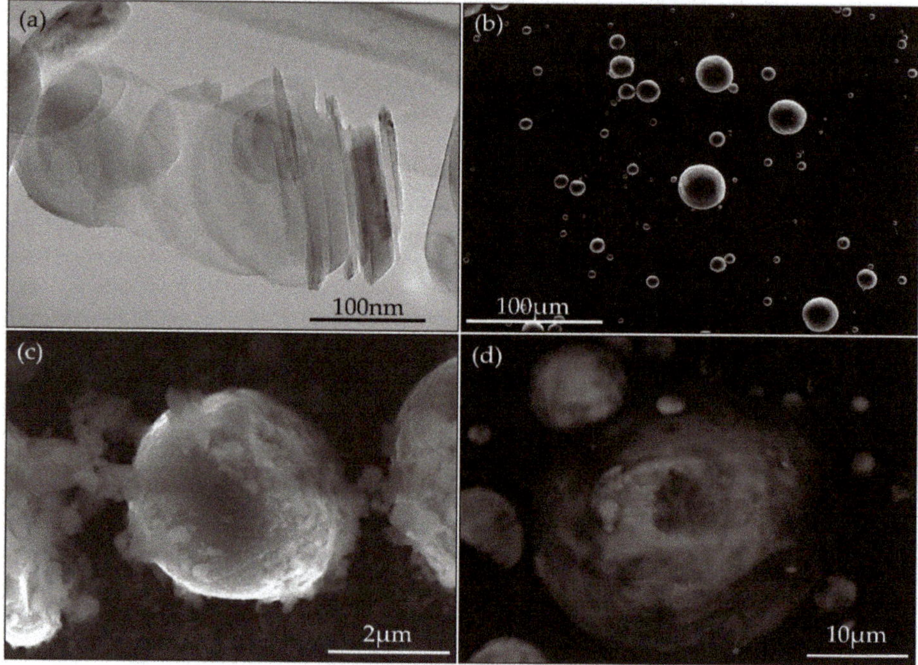

Figure 1. TEM and SEM images of powders before and after mixing: (**a**) TEM image of boron nitride powder as received; (**b**) SEM image of CP-Ti powder as received; (**c**) secondary electron SEM image of powders after mixing; (**d**) back-scatter electron image of powders after mixing.

3.2. Microstructure

Figure 2a is a low magnification TEM image showing a nanowhisker extracted from a sample sintered at 1050 °C for 6 h with an inset high magnification image. This particular nanowhisker had a length and diameter of approximately 18.4 μm and 47 nm, respectively, which was measured along with several others from each set of sintering conditions to obtain the average aspect ratios of the nanowhiskers in each sample. As shown in our previous study these whiskers are TiB [15]. However, to analyse the effect of BN concentration on the chemical nature of the samples, XRD was conducted on a CP-Ti + 10% BN powder sample and then the same powder after sintering at 1150 °C for 6 h with subsequent surface etching (Figure 2b). Each peak in the powder sample was indexed by either hexagonal BN or hexagonal αTi. The XRD peaks in the sintered sample were all indexed by orthogonally-structured TiB and αTi (Figure 2b), leading to two key findings. Firstly, no BN was detected in the samples after sintering, either in XRD or visible in SEM and TEM imaging, showing that all BN has undergone the in situ reaction. Secondly, TiN was also not detected in XRD or SEM and TEM imaging. TiN is regularly present as a product of the reaction between Ti and BN in rapid melting and solidification processes such as laser engineered net shaping (LENS) [25]. However, in slow heating and cooling solid-state processes, such as the furnace sintering employed in this work, no TiN is observed [34,35]. Nitrogen has a wide range of solubility in CP-Ti, >8 wt% at 1050 °C but approaching <0.1 wt% at room temperature [36]. Furnace cooling (2 °C/min) allows the sample to achieve the low temperature equilibrium state with very low solute nitrogen, which was not detected by XRD or EDS in this work or previous studies [15]. The combination of the low solubility of nitrogen at room temperature, the achievement of room temperature equilibrium in slow cooling processes and the absence of detected TiN leads to the conclusion that the nitrogen from BN in this work is lost to the atmosphere during processing.

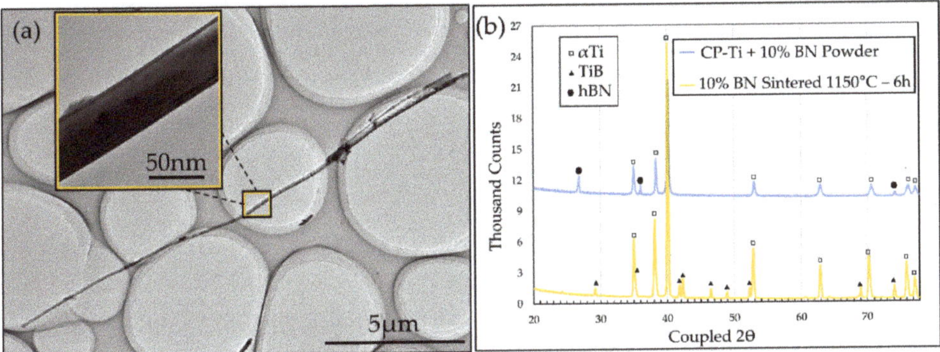

Figure 2. (a) TEM image showing a nanowhisker extracted from 10% BN sample after sintering at 1050 °C for 6 h, with inset high magnification image for measuring whisker diameter; (b) XRD pattern of 10% BN powder before sintering and after sintering at 1150 °C for 6 h.

In order to study the effect of BN concentration SEM images were taken at similar magnification for each of the sintering conditions used. Figure 3a shows a 5 vol% BN TMC sintered at 1200 °C for 6 h and etched for 30 s to reveal details of the microstructure around grain boundaries. For comparison, Figure 3b–d show 10 vol% BN TMCs sintered for 6 h at 1050 °C, 1150 °C and 1200 °C respectively, before 30 s etching. Each microstructure shows Ti grains with nano and microwhiskers of TiB. In order to investigate the effect of reinforcement, two batches of TMC powder were prepared, one with 5 vol% BN and one with 10 vol% BN. The 5 vol% BN sample sintered at 1200 °C includes mostly microwhiskers with average length of approximately 20 μm and diameter of 1 μm (see Figure 3a). Similarly, when the concentration of BN is doubled to 10 vol% with the same sintering conditions, TiB whiskers exhibit a similar structure with average length of approximately 18 μm and diameter of 0.8 μm (see Figure 3d). This comparison shows that up to 10 vol% of BN, the fraction of BN has little effect of the formation and growth of TiB.

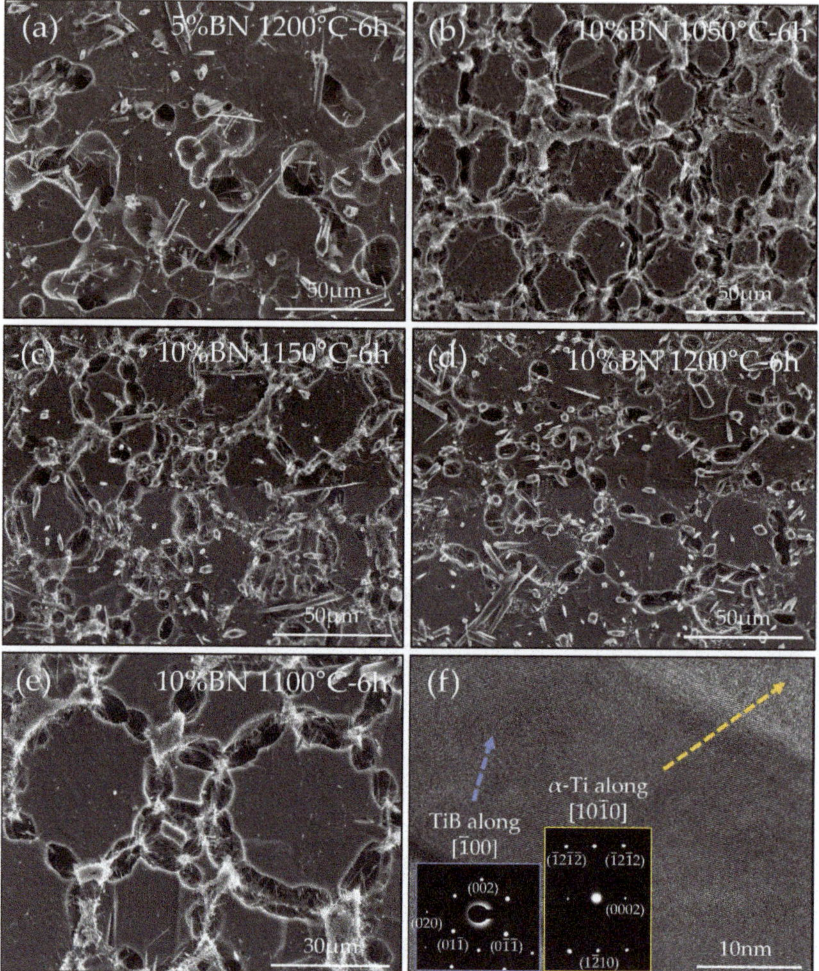

Figure 3. TEM and SEM images showing as-sintered TMC microstructure after etching: (**a**) SEM image of 5 vol% BN TMC sintered at 1200 °C for 6 h; (**b**) SEM image of 10 vol% BN TMC sintered at 1050 °C for 6 h; (**c**) SEM image of 10 vol% BN TMC sintered at 1150 °C for 6 h; (**d**) SEM image of 10 vol% BN TMC sintered at 1200 °C for 6 h; (**e**) SEM image of 10 vol% BN TMC sintered at 1100 °C for 6 h; (**f**) TEM image of FIB cut lamella from 10 vol% BN TMC sintered at 1100 °C for 6 h with inset indexed selected area electron diffraction patterns of TiB and α-Ti.

In addition to the effect of composition, the impact of processing temperature was investigated. Interestingly, when comparing Figure 3b–d a clear increase in the lateral dimension of TiB whiskers is observable. After sintering at 1050 °C, as in Figure 3b, there are very few TiB whiskers with a diameter greater than 100 nm. Despite significant clustering of TiB, there is a propensity for TiB to remain as nanowhiskers after processing at 1050 °C. TiB in the TMC sample sintered at 1050 °C for 6 h had an average diameter of 50 nm and length of 19 μm. By comparison, as temperature is increased to 1150 °C the number of TiB microwhiskers is increased, while still having significant amounts of TiB nanowhiskers (shown in Figure 3c). The TiB in the sample sintered at 1150 °C for 6 h has an average diameter of 65 nm and length of 23 μm. As temperature is further increased to 1200 °C, as shown in

Figure 3d, few TiB nanowhiskers are observed, with the majority coarsened to microwhiskers. It is clear that between 1150 °C and 1200 °C the coarsening of TiB becomes active. This temperature range is consistent with previous studies [14,15], which were conducted on TMCs with lower boron additions indicating that the fraction of TiB present, up to 10 vol%, in the sample has little effect on the critical coarsening temperature.

Figure 3e shows the microstructure of the TMCs produced by this method. After the removal of the Ti in the grain boundary a network structure of TiB interconnecting the Ti grains is observed. The use of low-energy mixing, fine BN and a solid-state sintering operation serves to preserve the structure. As TiB is formed in situ during sintering it grows from the surface of Ti particles into the adjacent particle acting as a grain boundary reinforcement. This microstructure is consistent with that observed in previous works [14,15,37–39]. The TiB present in the grain boundary also serves to restrict the grain growth of Ti grains as is commonly observed in slow heating/cooling processes like furnace sintering [37]. The average grain size in each 10 vol% BN sample was approximately 7 μm. A primary detriment to the furnace sintering approach is the coarsening of the grain structure. Therefore, the formation of TMCs by this method, which restricts grain size to a minimum, is of strong benefit.

Figure 3f and insets show a TEM image and selected area electron diffraction (SAED) patterns. Here, a section of a TiB nanowhisker with (011) atomic planes can be observed, indicating that the TiB nanowhisker is a single crystal. The provided SAED patterns provide evidence that the nanowhiskers are TiB with the typical orthorhombic structure, while Ti is present in the HCP structured alpha phase. As expected, no β-Ti was observed in the sample, only α-Ti, due to the slow cooling rate (2 °C/min) allowing for the sample to achieve the equilibrium alpha phase. This thermodynamically stable microstructure is beneficial for long service bone implants as there is less risk of microstructural and therefore property changes over time [40].

3.3. Mechanical Properties

α-Ti has excellent corrosion resistance and promotes rapid osteointegration because of a stable TiO_2 film, where OH^- ions present in this surface layer react with ions present in the bone structure like Ca^{2+} and PO_4^{3-} to aid adhesion [7]. These properties make α-Ti a strong candidate for bone implants. However, the poor fatigue performance and overall low strength have limited the application of CP-Ti and other α-Ti alloys in load bearing implants [7]. A TMC with an α-Ti matrix may offer improved strength and wear properties whilst retaining the corrosion resistance and biological interfacing benefits of α-Ti.

Table 1 presents both results from nanoindentation tests along with mechanical property results from other works. When calculating average E and H, values outside two standard deviations above the mean were rejected. This step was taken as surface damage and contamination can lead to large errors in results when performing nanoindentation [41]. Table 1 shows that our TMC samples exhibit high E and H values when compared to both CP-Ti and other TMCs, even when a higher reinforcement fraction is used. Most notably, the H/E ratio of samples is very high. The ratio of hardness to modulus is directly related to wear resistance and strain to failure [42]. The 10 vol% BN sample sintered at 1150 °C for 6 h has the highest H/E ratio of 0.062, almost twice that of CP-Ti. The ratio H^3/E^2, called yield pressure, is associated with yield strength [42,43]. The 10 vol% BN sample sintered at 1150 °C for 6 h has the highest H^3/E^2 ratio (0.041 GPa), which is an order of magnitude higher than that of CP-Ti and significantly higher than reported TMCs.

There are several contributing factors to the high performance of our TMCs, one of which is the previously described formation of a network microstructure with reduced grain size. As grain size is reduced, TiB grown in the grain boundary is sufficiently long to grow across the Ti grains, as can be seen in Figure 3. This leads to the improved local property measurement by nanoindentation throughout Ti grains, due to the strengthening effect of TiB. In addition to the increase in hardness and elastic modulus observed, there is an expected increase in bulk material strength as per the Hall–Petch relationship due to reduced grain size [44]. Also, network microstructures have previously been shown to improve both strength and ductility of TMCs [15,21,38,45,46]. Another key factor contributing to the performance of our TMCs is the high aspect ratio of TiB nanowhiskers, particularly in 10 vol% BN samples. The aspect ratio of TiB in samples sintered at 1150 °C and lower is approximately 400. It is well understood that increasing the aspect ratio of the reinforcement phase increases the strengthening efficiency of the reinforcement [23]. However, typically, when the fraction of reinforcement is increased, clustering of the secondary particles prior to the in situ reaction leads to coarsening of the reinforcement phase and, therefore, a decreasing of the strengthening efficiency [33]. In this work, BN fraction showed no effect on the size of TiB, with high aspect ratio TiB nanowhiskers present when 10 vol% BN was used. Figure 3b–d show several fine TiB nanowhiskers grouped together which had not coarsened to microwhiskers. Previous studies have shown that there is a critical temperature for the coarsening of TiB between 1150 and 1200 °C with low TiB fraction (less than 5 vol%) [14,15]. It is clear from this study that TiB retains this critical temperature for coarsening when volume fraction is increased to 10 vol%. Analysis of the formation reaction and coarsening of TiB in detail through in situ microscopy studies may provide necessary information to assist with the design of optimum manufacturing processes, both in solid and melt methods. The combination of increased volume fraction of TiB and maintaining ultrahigh aspect ratio nanowhiskers plays an important role in the enhanced properties of our TMCs.

Table 1. Summary of key results from nanoindentation tests, with results from other works for comparison.

Sample Composition	Process Parameters	Hardness/St. Dev (GPa)	Elastic Modulus/St. Dev (GPa)	H/E	H^3/E^2 (GPa)	Reference
CP-Ti	1200 °C—6 h	3.43/0.29	94.4/2.3	0.036	0.005	This work
CP-Ti	SLM 120 J/mm^2	2.39	102	0.023	0.001	[47]
Ti + 5 vol% BN	1200 °C—6 h	7.50/0.70	147.7/15.4	0.051	0.020	This work
Ti + 10 vol% BN	1050 °C—6 h	10.01/1.42	170.7/21.0	0.059	0.034	This work
Ti + 10 vol% BN	1150 °C—6 h	10.48/1.23	167.9/17.2	0.062	0.041	This work
Ti + 5 vol% TiB$_2$	SLM 120 J/mm^2	3.33	122	0.027	0.003	[47]
Ti + 5 vol% TiB$_2$	SPS 1150 °C—5 min	4.1	-	-		[48]
Ti + 5 vol% TiB$_2$	SPS 1250 °C—5 min	4.3	-	-		
Ti + 25 vol% TiB$_2$	SPS 1150 °C—5 min	7.1	-	-		
Ti + 25 vol% TiB$_2$	SPS 1250 °C—5 min	10.1	-	-		
Ti + 24 vol%TiB	SPS 1100 °C—5 min	8.2 [1]	162.6	0.050	0.021	[49]
Ti + 38.5 vol%TiB	1200 °C—5 h	9.7 [1]	190.5	0.051	0.025	
Ti + 20.6 vol%TiB	HIP 1200 °C, 120 MPa—5 h	8.0 [1]	169.5	0.047	0.018	

[1] Originally reported as Vickers hardness, converted to GPa.

4. Conclusions

To summarize, a TiB nanowhisker-reinforced α-Ti matrix TMC was produced using a method of low energy mixing, cold compaction and furnace sintering. This TMC demonstrated increased strength and wear performance, addressing the two key weaknesses presently facing the use of α-Ti for biomedical implants. Overcoming these challenges is a critical improvement for the development of affordable long-lasting biomedical implants to reduce the number of replacement surgeries required. Analysis showed that TiB is present as nanowhiskers with high aspect ratio in samples sintered at 1150 °C or below with relatively high volume fraction, up to 10 vol% BN. TiB nanowhiskers were observed to coarsen only when the sintering temperature was increased to 1200 °C, even in high volume fractions of BN, shedding light on the in situ reaction process and how it can be manipulated to achieve a refined reinforcement phase. The combination of high aspect ratio TiB in high volume fraction, a fine grain α-Ti matrix and a network microstructure led to an increase in hardness and reduced modulus of more than 300 and 170% respectively when compared to CP-Ti. By improving these key properties while retaining the strong osteointegration and corrosion resistance of an α-Ti matrix these TMCs are a strong candidate for the future of long-lasting load bearing implants. The TMCs produced in this work progress towards addressing the current challenges facing the biomedical implant industry by providing insight into: TiB reinforced TMCs, the in situ reactions at play and their optimized manufacture. Materials such as these will be vital to the next generation of high-performance load-bearing biomaterials.

Author Contributions: Conceptualization: J.A.O., M.S.D. and J.Z.; methodology: J.A.O. and M.S.D.; software: J.A.O. and R.P.; validation: J.A.O. and R.P.; formal analysis: J.A.O. and R.P.; investigation: J.A.O. and R.P.; resources: M.S.D., M.L. and J.Z.; data curation: J.A.O. and R.P.; writing—original draft preparation: J.A.O.; writing—review and editing: J.A.O., M.S.D. and J.Z.; visualization: J.A.O.; supervision: M.S.D., M.L. and J.Z.; project administration: J.A.O., M.S.D.; funding acquisition: M.S.D. All authors have read and agreed to the published version of the manuscript.

Funding: This research was funded by the University of Queensland's Centre for Advanced Materials Processing and Manufacturing (AMPAM) by the ARC research hub IH150100024.

Acknowledgments: The authors acknowledge the facilities, and the scientific and technical assistance, of the Microscopy Australia Facility at the Centre for Microscopy and Microanalysis, The University of Queensland.

Conflicts of Interest: The authors declare no conflict of interest.

References

1. Geetha, M.; Singh, A.K.; Asokamani, R.; Gogia, A.K. Ti based biomaterials, the ultimate choice for orthopaedic implants—A review. *Prog. Mater. Sci.* **2009**, *54*, 397–425. [CrossRef]

2. Long, M.; Rack, H.J. Titanium alloys in total joint replacement—A materials science perspective. *Biomaterials* **1998**, *19*, 1621–1639. [CrossRef]

3. Kurtz, S.; Ong, K.; Lau, E.; Mowat, F.; Halpern, M. Projections of primary and revision hip and knee arthroplasty in the United States from 2005 to 2030. *J. Bone Jt. Surg. Am. Vol.* **2007**, *89*, 780–785. [CrossRef]

4. Zhang, L.-C.; Chen, L.-Y. A Review on Biomedical Titanium Alloys: Recent Progress and Prospect. *Adv. Eng. Mater.* **2019**, *21*, 1801215. [CrossRef]

5. Chen, Y.; Frith, J.E.; Dehghan-Manshadi, A.; Attar, H.; Kent, D.; Soro, N.D.M.; Bermingham, M.J.; Dargusch, M.S. Mechanical properties and biocompatibility of porous titanium scaffolds for bone tissue engineering. *J. Mech. Behav. Biomed. Mater.* **2017**, *75*, 169–174. [CrossRef]

6. Kurtz, S.M.; Ong, K.L.; Schmier, J.; Mowat, F.; Saleh, K.; Dybvik, E.; Karrholm, J.; Garellick, G.; Havelin, L.I.; Furnes, O.; et al. Future clinical and economic impact of revision total hip and knee arthroplasty. *J. Bone Jt. Surg. Am. Vol.* **2007**, *89A*, 144–151. [CrossRef]

7. Mitragotri, S.; Lahann, J. Physical approaches to biomaterial design. *Nat. Mater.* **2009**, *8*, 15–23. [CrossRef]

8. Besra, L.; Liu, M. A review on fundamentals and applications of electrophoretic deposition (EPD). *Prog. Mater. Sci.* **2007**, *52*, 1–61. [CrossRef]

9. Boyd, A.R.; Rutledge, L.; Randolph, L.D.; Meenan, B.J. Strontium-substituted hydroxyapatite coatings deposited via a co-deposition sputter technique. *Mater. Sci. Eng. C Mater. Biol. Appl.* **2015**, *46*, 290–300. [CrossRef]

10. Liu, X.Y.; Ding, C.X.; Wang, Z.Y. Apatite formed on the surface of plasma-sprayed wollastonite coating immersed in simulated body fluid. *Biomaterials* **2001**, *22*, 2007–2012. [CrossRef]

11. Paital, S.R.; Dahotre, N.B. Calcium phosphate coatings for bio-implant applications: Materials, performance factors, and methodologies. *Mater. Sci. Eng. R Rep.* **2009**, *66*, 1–70. [CrossRef]

12. Egbo, M.K. A fundamental review on composite materials and some of their applications in biomedical engineering. *J. King Saud Univ. Eng. Sci.* **2020**. [CrossRef]

13. Feng, H.; Zhou, Y.; Jia, D.; Meng, Q. Microstructure and mechanical properties of in situ TiB reinforced titanium matrix composites based on Ti–FeMo–B prepared by spark plasma sintering. *Compos. Sci. Technol.* **2004**, *64*, 2495–2500. [CrossRef]

14. Huang, L.; Qian, M.; Liu, Z.; Nguyen, V.T.; Yang, L.; Wang, L.; Zou, J. In situ preparation of TiB nanowires for high-performance Ti metal matrix nanocomposites. *J. Alloy. Compd.* **2018**, *735*, 2640–2645. [CrossRef]

15. Otte, J.A.; Zou, J.; Huang, Y.; Dargusch, M.S. Ultrahigh Aspect Ratio TiB Nanowhisker-Reinforced Titanium Matrix Composites as Lightweight and Low-Cost Replacements for Superalloys. *Acs Appl. Nano Mater.* **2020**, *3*, 8208–8215. [CrossRef]

16. Ravi Chandran, K.; Panda, K.; Sahay, S. TiB Nanowhisker-reinforced Ti composites: Processing, properties, application prospects, and research needs. *J. Miner. Met. Mater. Soc. (TMS)* **2004**, *56*, 42–48. [CrossRef]

17. Saito, T.; Takamiya, H.; Furuta, T. Thermomechanical properties of P/M β titanium metal matrix composite. *Mater. Sci. Eng. A* **1998**, *243*, 273–278. [CrossRef]

18. Tamirisakandala, S.; Bhat, R.B.; Ravi, V.A.; Miracle, D.B. Powder metallurgy Ti-6Al-4V-xB alloys: Processing, microstructure, and properties. *JOM* **2004**, *56*, 60–63. [CrossRef]

19. Tao, X.; Yao, Z.; Zhang, S. Reconstruction and refinement of TiB whiskers in titanium matrix composite after electron beam remelting. *Mater. Lett.* **2018**, *225*, 13–16. [CrossRef]

20. Kondoh, K. 16—Titanium metal matrix composites by powder metallurgy (PM) routes. In *Titanium Powder Metallurgy*; Qian, M., Froes, F.H., Eds.; Butterworth-Heinemann: Boston, MA, USA, 2015; pp. 277–297. [CrossRef]

21. Lu, K. The Future of Metals. *Science* **2010**, *328*, 319. [CrossRef]

22. Zhang, C.J.; Kong, F.T.; Xiao, S.L.; Zhao, E.T.; Xu, L.J.; Chen, Y.Y. Evolution of microstructure and tensile properties of in situ titanium matrix composites with volume fraction of (TiB+TiC) reinforcements. *Mater. Sci. Eng. A* **2012**, *548*, 152–160. [CrossRef]

23. Koo, M.Y.; Park, J.S.; Park, M.K.; Kim, K.T.; Hong, S.H. Effect of aspect ratios of in situ formed TiB whiskers on the mechanical properties of TiBw/Ti–6Al–4V composites. *Scr. Mater.* **2012**, *66*, 487–490. [CrossRef]

24. Makau, F.M.; Morsi, K.; Gude, N.; Alvarez, R.; Sussman, M.; May-Newman, K. Viability of Titanium-Titanium Boride Composite as a Biomaterial. *ISRN Biomater.* **2013**, *2013*, 970535. [CrossRef]

25. Das, M.; Bhattacharya, K.; Dittrick, S.A.; Mandal, C.; Balla, V.K.; Sampath Kumar, T.S.; Bandyopadhyay, A.; Manna, I. In situ synthesized TiB–TiN reinforced Ti6Al4V alloy composite coatings: Microstructure, tribological and in-vitro biocompatibility. *J. Mech. Behav. Biomed. Mater.* **2014**, *29*, 259–271. [CrossRef]

26. Kaczmarek, M.; Jurczyk, M.U.; Miklaszewski, A.; Paszel-Jaworska, A.; Romaniuk, A.; Lipińska, N.; Żurawski, J.; Urbaniak, P.; Jurczyk, K. In vitro biocompatibility of titanium after plasma surface alloying with boron. *Mater. Sci. Eng. C* **2016**, *69*, 1240–1247. [CrossRef]

27. Chen, Y.; Zhang, J.; Dai, N.; Qin, P.; Attar, H.; Zhang, L.-C. Corrosion Behaviour of Selective Laser Melted Ti-TiB Biocomposite in Simulated Body Fluid. *Electrochim. Acta* **2017**, *232*, 89–97. [CrossRef]

28. Morsi, K. Review: Titanium–titanium boride composites. *J. Mater. Sci.* **2019**, *54*, 6753–6771. [CrossRef]

29. Ege, D.; Duru, İ.; Kamali, A.R.; Boccaccini, A.R. Nitride, Zirconia, Alumina, and Carbide Coatings on Ti6Al4V Femoral Heads: Effect of Deposition Techniques on Mechanical and Tribological Properties. *Adv. Eng. Mater.* **2017**, *19*, 1700177. [CrossRef]

30. Giannuzzi, L.A.; Stevie, F.A. A review of focused ion beam milling techniques for TEM specimen preparation. *Micron* **1999**, *30*, 197–204. [CrossRef]

31. Oliver, W.C.; Pharr, G.M. An improved technique for determining hardness and elastic modulus using load and displacement sensing indentation experiments. *J. Mater. Res.* **1992**, *7*, 1564–1583. [CrossRef]

32. Fang, Z.Z.; Paramore, J.D.; Sun, P.; Chandran, K.S.R.; Zhang, Y.; Xia, Y.; Cao, F.; Koopman, M.; Free, M. Powder metallurgy of titanium—Past, present, and future. *Int. Mater. Rev.* **2018**, *63*, 407–459. [CrossRef]

33. Brown, M.E.; Dollimore, D.; Galwey, A.K. Reactions in the Solid State. In *Comprehensive Chemical Kinetics*; Bamford, C.H., Tipper, C.F.H., Eds.; Elsevier: Amsterdam, The Netherlands, 1980; Volume 22.

34. Bhuiyan, M.M.H.; Li, L.H.; Wang, J.; Hodgson, P.; Chen, Y. Interfacial reactions between titanium and boron nitride nanotubes. *Scr. Mater.* **2017**, *127*, 108–112. [CrossRef]

35. Huang, L.; Qian, M.; Wang, L.; Chen, Z.-G.; Shi, Z.; Nguyen, V.; Zou, J. High-tensile-strength and ductile novel Ti-Fe-N-B alloys reinforced with TiB nanowires. *Mater. Sci. Eng. A* **2017**, *708*, 285–290. [CrossRef]

36. Wriedt, H.A.; Murray, J.L. The N-Ti (Nitrogen-Titanium) system. *Bull. Alloy Phase Diagr.* **1987**, *8*, 378–388. [CrossRef]

37. Huang, L.J.; Geng, L.; Peng, H.X. In situ (TiBw+TiCp)/Ti6Al4V composites with a network reinforcement distribution. *Mater. Sci. Eng. A* **2010**, *527*, 6723–6727. [CrossRef]

38. Huang, L.J.; Wang, S.; Dong, Y.S.; Zhang, Y.Z.; Pan, F.; Geng, L.; Peng, H.X. Tailoring a novel network reinforcement architecture exploiting superior tensile properties of in situ TiBw/Ti composites. *Mater. Sci. Eng. A* **2012**, *545*, 187–193. [CrossRef]

39. Wang, S.; Huang, L.; An, Q.; Geng, L.; Liu, B. Dramatically enhanced impact toughness of two-scale laminate-network structured composites. *Mater. Des.* **2018**, *140*, 163–171. [CrossRef]

40. Oshida, Y. 5—Mechanical and Tribological Behaviors. In *Bioscience and Bioengineering of Titanium Materials*, 2nd ed.; Oshida, Y., Ed.; Elsevier: Oxford, UK, 2013; pp. 117–137. [CrossRef]

41. Dao, M.; Chollacoop, N.; Van Vliet, K.J.; Venkatesh, T.A.; Suresh, S. Computational modeling of the forward and reverse problems in instrumented sharp indentation. *Acta Mater.* **2001**, *49*, 3899–3918. [CrossRef]

42. Xu, J.; Wang, G.D.; Lu, X.; Liu, L.; Munroe, P.; Xie, Z.-H. Mechanical and corrosion-resistant properties of Ti–Nb–Si–N nanocomposite films prepared by a double glow discharge plasma technique. *Ceram. Int.* **2014**, *40*, 8621–8630. [CrossRef]

43. Fornell, J.; Van Steenberge, N.; Varea, A.; Rossinyol, E.; Pellicer, E.; Suriñach, S.; Baró, M.D.; Sort, J. Enhanced mechanical properties and in vitro corrosion behavior of amorphous and devitrified Ti40Zr10Cu38Pd12 metallic glass. *J. Mech. Behav. Biomed. Mater.* **2011**, *4*, 1709–1717. [CrossRef]

44. Hall, E.O. The deformation and ageing of mild steel: Iii discussion of results. *Proc. Phys. Soc. Sect. B* **1951**, *64*, 747–753. [CrossRef]

45. Hu, Y.; Ning, F.; Wang, H.; Cong, W.; Zhao, B. Laser engineered net shaping of quasi-continuous network microstructural TiB reinforced titanium matrix bulk composites: Microstructure and wear performance. *Opt. Laser Technol.* **2018**, *99*, 174–183. [CrossRef]

46. Huang, L.J.; Geng, L.; Peng, H.X. Microstructurally inhomogeneous composites: Is a homogeneous reinforcement distribution optimal? *Prog. Mater. Sci.* **2015**, *71*, 93–168. [CrossRef]

47. Attar, H.; Ehtemam-Haghighi, S.; Kent, D.; Okulov, I.V.; Wendrock, H.; Bönisch, M.; Volegov, A.S.; Calin, M.; Eckert, J.; Dargusch, M.S. Nanoindentation and wear properties of Ti and Ti-TiB composite materials produced by selective laser melting. *Mater. Sci. Eng. A* **2017**, *688*, 20–26. [CrossRef]

48. Wei, S.; Zhang, Z.-H.; Wang, F.-C.; Shen, X.-B.; Cai, H.-N.; Lee, S.-K.; Wang, L. Effect of Ti content and sintering temperature on the microstructures and mechanical properties of TiB reinforced titanium composites synthesized by SPS process. *Mater. Sci. Eng. A* **2013**, *560*, 249–255. [CrossRef]

49. Selvakumar, M.; Chandrasekar, P.; Ravisankar, B.; Balaraju, J.; Mohanraj, M. Mechanical Properties of Titanium–Titanium Boride Composites Through Nanoindentation and Ultrasonic Techniques—An Evaluation Perspective. *Powder Metall. Met. Ceram.* **2015**, *53*, 557–565. [CrossRef]

Publisher's Note: MDPI stays neutral with regard to jurisdictional claims in published maps and institutional affiliations.

MDPI
St. Alban-Anlage 66
4052 Basel
Switzerland
Tel. +41 61 683 77 34
Fax +41 61 302 89 18
www.mdpi.com

Nanomaterials Editorial Office
E-mail: nanomaterials@mdpi.com
www.mdpi.com/journal/nanomaterials